The Seven Cs of Stress

A Burkean Approach

Stan A. Lindsay

Published by

Say Press
Orlando, FL

Library of Congress Control Number: 2004099074
ISBN: 978-0-9841491-8-6

Contents

To Shane, Chari, Auburn, and Tristan

Foreword (by Sandra Halvorson)

My doctoral dissertation pertained to stress. I have conducted seminars on stress management, as a business consultant. I teach courses in conflict resolution at the graduate and undergraduate levels. And, like all readers, I have experienced my share of stress in life. In my opinion, Stan A. Lindsay has done an outstanding job of combining his experiences in business and academia and his life experiences into a work that showcases the pertinent contributions of Kenneth Burke, Aristotle, and others in the field of communication. Lindsay skillfully guides the reader through a refreshingly different approach to dealing with stress. His heart-warming stories are easy to relate to as the reader undertakes the journey. He offers an in-depth look at the concept of stress and suggests step-by-step solutions (which he identifies as "relief valves") to aid the reader in successfully managing stress. Lindsay metaphorically takes the reader through seven Cs of stress much as a captain successfully navigates a voyage at sea. He groups the stressors into seven categories of stress: Corporal (stress of the body), Community (stress experienced when dealing with other people), Cash (stress concerning how to handle money wisely), Chrono (stress dealing with managing time), Competence (stress dealing with questions about one's ability to perform a task), Confusion (stress pertaining to decision-making and other situations in which one feels lost), and Conscientious (stress concerning morality).

I recently managed to experience in one short hour all seven of the stressors that Lindsay presents in his book. I had successfully managed to endure the sound of the 5:00 a.m. wakeup call from my demonic alarm clock earlier in the morning. "Triumph" was the word that came to mind as I successfully dislodged my newspaper from the thorny bush located at the end of my driveway. I was a little discouraged when my high-protein, low-calorie, low-fat package of whole grain morsels tipped upside down and avalanched into the tangled web of cords at the base of my computer. I know that caffeine, sugar, and maltrodexin are bad for you but I was really looking forward to supplementing my healthy snack with a low-fat coffee drink. Never mind. The clerk inadvertently forgot to put it in my grocery bag. I was beginning to get a little uptight when I turned on my computer and suspicious sounds began to erupt from the bowels of the CPU. The sounds were quite similar to those made by my car just before the mechanic said, "You have two options, lady—either fork up twenty-five hundred to fix it or buy a new car." As the groans continued, I noticed that it was cold in the house. I immediately inspected the newly installed thermostat. When I delicately pressed the central button, there was no immediate response. When I defiantly pushed all the buttons, there was a response—absolutely no heat. By 6:00 a.m., I was sitting at my ergonomically incorrect workstation at my computer that I have lovingly nicknamed, "Old Yeller." I was sitting on the edge of my skewed chair, feet precariously placed on a lower shelf in an attempt to avoid getting any more crunchy bits of wholegrain globs of nutrition on the bottom of my feet. My teeth were chattering because of the cold. I had little faith that a service person would eagerly be available to repair my thermostat at the wee hours of a Saturday morning.

In that hour, I was exposed to Corporal stress (as a result of having too little sleep and an inadequate configuration for a workstation). I found myself becoming a bit grumpy with and aching back. Community stress resulted when I experienced feelings of resentment towards the delivery person and the stock clerk. Cash stress came as I anticipated the cost of purchasing a new computer and/or a new thermostat. Chrono stress arrived as I wondered when I would have time to grade papers, go shopping for Christmas specials, and still have time to arrang for that repairperson. My inability to fix the thermostat produced thoughts about Competence stress: Whom should I call? Where did I put that warranty?

Confusion stress. My thoughts were turning to language that I suppressed due to Conscientious stress.

Although the stressors I navigated in that hour were relatively mild, an accumulation of multiple stresses over time, without the exercise of appropriated relief valves, leads to damaging and sometimes fatal health conditions. *The American Heritage Dictionary* defines stress: "A state of extreme difficulty, pressure, or strain. An applied force or system of forces that tends to strain or deform a body." Although this definition refers more to deforming a metal object, it also carries implications related to the human body. Humans are faced with stressful situations every day. The stress may pertain to major woes such as terrorism or the loss of a loved one. Some stress is even desirable, such as that which comes from athletic competition and viewing scary movies. However, when eustress becomes distress, stressors often take a toll on the human body, manifesting themselves in the form of heart disease, liver disease, lung disease, cancer, accidents, immune system deficiency, memory loss, obesity, etc.

Certainly, few health topics appear to match the appeal of "stress-related" books. Recent software regarding books in print lists 1,985 titles dealing with "stress." According to Amazon.com, 6,701 book titles dealing with "stress" are available. Of these, several are self-help books, but only a few titles deal with the "sources of stress/stressors." This book identifies the seven specific stressors and supplies solutions that readers can apply now.

Some of the titles that deal with stressors are quite specialized, such as Kathleen Heide's 1992 offering, *Why Kids Kill Parents: Child Abuse and Adolescent Homicide*, Robert Handly's book (first published in 1985), *Anxiety and Panic Attacks: Their Cause and Cure)*, and W. David Hager, M.D. and Linda Carruth Hager's 1995 work, *Stress and the Woman's Body*. Handly's book appears to have staying power but it is almost a workbook in its step-by-step approach. The Hagers' book is a Christian approach and is an expansion on what Lindsay calls "corporal stress."

Other works, such as Susan McMahon's 1996 work, *The Portable Problem Solver*, are for general audiences. Long-lasting titles like Edward A. Charlesworth and Ronald G. Nathan's *Stress Management* (1982, 1985), Christopher J. McCullough and Robert Woods Mann's *Managing Your Anxiety: Regaining Control When You Feel Stressed, Helpless, and Alone* (1986, 1994), and Ronald Nathan, Thomas E. Staats, and Paul J. Rosch's *The Doctor's Guide to Instant*

Stress Relief often are thoroughly clinical in their approach. They read like textbooks and provide many relief bandaids. These "bandaid" approaches deal primarily with the symptoms of stress, not the causes. *The Seven C's of Stress* deals with the causes of stress, and proposes solutions based upon the individual causes.

The vast majority of the texts available are from authors with psychological or medical credentials. *The Seven C's of Stress*, on the other hand, is written by a person with "communication" credentials. The approach is different. *The Seven C's of Stress* stands alone in identifying the seven specific sources/stressors: corporal, cash, community, confusion, competence, conscientious, and chrono-stress. It does not read like a medical treatise (as do most of its competitors). It is easy-to-read because it incorporates real life examples, representative anecdotes-- the stresses faced by real people in real situations. The primary audience for *The Seven C's of Stress* is those interested in dealing with stress in business. However, it is also valuable for those interested in the stresses of relationships, marriage and family.

Sandra Halvorson, Ph.D.

Acknowledgments

While I attribute the primary approach I make here to Kenneth Burke, I acknowledge that the seeds for this study were planted by Professor Clyde Narramore of the California Graduate School of Theology. Professor Narramore taught a course on stress management at the graduate school in the Summer of 1991. My original notes on stress from his course emphasized the notion of dealing with stress by determining the stressor. Althought the stressors developed in this book are more comprehensive than those identified in Dr. Narramore's course, the principle of identifying the stressor before developing a management strategy is Dr. Narramore's. I deeply thank him, first of all, for his role in generating this book.

A second thank you goes to many of my undergraduate students in the English and Communication departments at Purdue University from 1991 to 2001. As I taught my courses, I presented my developing list of stressors to each section. I asked students if they could think of any stressors I may have missed. I also asked my students what relief valves they used to lessen each type of stress. Their feedback was very helpful. Ultimately, I determined that there are seven major categories of stressors. I found that each category could be labeled with a word beginning with the letter C, and I saw my approach to stress as "sailing the seven Cs."

Although I have allowed my acquaintances to remain anonymous in this book, many of my examples, used here to illustrate stressful situations, are anecdotal. Individuals may question whether they are the ones who are referred to in any given case, but I have intentionally refrained from specific identifications. Nevertheless, I thank those who have provided examples of stress for the impact they had on my life and my understanding of stress-related phenomena.

Finally, I want to thank my colleague at Florida State University, Sandra Halvorson, Ph.D., for her willingness to write the Foreword to this book. In addition to her college teaching, Dr. Halvorson has a great deal of experience as a consultant, conducting stress management seminars for several different companies. Her take on my approach to stress management is quite insightful, and deeply appreciated.

Stan A. Lindsay

Chapter 1

Introduction

Businesspersons encounter a fair amount of stress. Virtually all human beings do. Some have the mistaken notion that it is possible or, at least, desirable to live life stress-free. They blame stress for producing ulcers, high blood pressure, and heart attacks (which stress does produce). On the other hand, life without stress would be quite boring. If there were no problems to solve, no peace to be negotiated, no mountains to climb, human beings would not be happy.

We are all goaded by one sense or another of climbing some ladder of success.[1] Our various definitions of success provide different labels for the various ladders, which might be climbed. One ladder is the ladder of wealth. Another is fame; another, social status; another, education. For some, the ladder is power: How much influence can we exert on other people? Even religious institutions have hierarchies. Indeed, the etymology of the word hierarchy is religious. It refers to the various levels of priesthood, which one might hold in the church. One might be a pope, a cardinal, an archbishop, a bishop, a priest, *etc*. The spirit of hierarchy

---

[1] The humanist philosopher, Kenneth Burke, lists this drive as one of the basic elements in his definition of human. Kenneth Burke, *Language as Symbolic Action* (Berkeley: Univ. of California Press, 1966), 15-16.

is not limited to any specific classification of humans. A little league baseball player wants to prove that s/he could play in the infield, instead of right field. A nursing home resident wants to be the champion checkers player at the nursing home. A teenage drummer wants his/her band to be offered a recording contract. A college student wants to graduate *summa cum laude.* An inner city youth wants to wear the most impressive athletic shoes. A Middle East dictator wants to demonstrate that he can stand up to a world super power. The list of hierarchies is endless. There are hierarchies among women, children, slaves, prisoners, laborers, and aristocrats. No humans are immune to the feeling of dissatisfaction when there are no more goals to be achieved. Hence, all humans crave stress.

Kenneth Burke claims that the body has a "natural aptitude for the undergoing of stress."[2] Burke phrases it poetically:

> If, to seek its level,
> Water can all the time
> Descend
> What God or Devil
> Makes men climb
> No end?[3]

Burke offers this poetry in the context of a late discussion of a clause in his definition of human--"goaded by the spirit of hierarchy."[4] Businesspersons would find life particularly unfulfilling, if it were not for this desire to climb a hierarchy. It is essentially a craving for the adventure, which produces stress that causes a person to leave the employ of someone else in order to succeed in his/her own business. Business "competition," like athletic competition, allows the businessperson to strategically flex managerial or economic "muscles," to welcome adversity. The businessperson has a love for the fight, for the opportunity to prove that s/he is capable of taking on all comers and emerging victorious.

Burke claims that this "spirit of hierarchy" is present in all humans. Stress specialist, Samuel Z. Klausner, comments:

---

[2]Kenneth Burke, "On Stress, Its Seeking," in *Why Man Takes Chances: Studies in Stress-Seeking,* ed. Samuel Z. Klausner (Garden City, NY: Doubleday, 1968), 99.
[3]Kenneth Burke, *The Rhetoric of Religion: Studies in Logology* (Berkeley: Univ. of California Press, 1970), 42.
[4]Burke, *Religion*, 40

> Kenneth Burke's stress-seeker is a divided self, troubled by necessity. He "must" create his own difficulties. Paradoxically, man struggles toward the enjoyment of Nirvana; yet to relax in Nirvana would be to allow his muscles to atrophy, and he would lose his ability to enjoy it. Thousands of years of adaptation have produced the need to seek stress. But more, in seeking stress he must risk that which he really possesses.[5]

No one understands the stress involved in "risk[ing] that which [one] . . . possesses" more than businesspersons.

Businesspersons may even be engaged in the process of bringing temporary stress to potential customers. As I discuss in my book, *Persuasion, Proposals, and Public Speaking* (Say Press, 2004), it is necessary for a salesperson to persuade the prospect that a problem exists before providing the solution. During the time, which intervenes between the persuasion that a problem exists and the enactment of the solution, the prospect experiences some stress. Burke includes "a reference to a practice of some current pitchmen, who help reinforce your sense of guilt by selling you remedies after first persuading you that you need them."[6] As an example, Burke remarks that advertisers point out to us that our bodies stink. Therefore, we need to gargle with poison (*i.e.*, mouthwash) or clog the pores of our skin (*i.e.*, with deodorants) to minimize the odor. The solution is designed to relieve the stress, which the advertisement has produced.

Whether stress is self-induced (as when businesspersons take risks) or induced by other persons (as when advertisers persuade customers that a problem exists), stress functions in fairly predictable ways. For that matter, stress operates uniformly whether the stress comes from another person, from our own bodies, from a financial imbalance, from a confusing situation, from a moral code, or from any other source.

Stress is like a balloon. When one inflates the balloon enough to stretch its membrane to the point at which it resembles a sphere, oval, or whatever shape it is designed to have, it loses that limpness which it originally possessed. This is stress. It is unnecessary to inflate the balloon to its full capacity in order to observe stress. It is unnecessary to have even one pound of pressure in the balloon

---

[5]Samuel Z. Klausner, ed., *Why Man Takes Chances: Studies in Stress-Seeking* (Garden City, NY: Doubleday, 1968), 74.
[6]Burke, "On Stress, Its Seeking," 100.

in order to observe stress. In fact, the same amount of pressure that it takes to inflate a paper bag would still demonstrate the existence of observable stress.

Eustress

The level of stress that is observed in an inflated paper bag or in an equally inflated balloon is actually a "good" type of stress. In the balloon example, even though the balloon is inflated, the rubber membrane has so much "give" or "flexibility" that you can actually squeeze it firmly without the balloon bursting. You can poke your finger into the balloon from all directions, and the balloon will survive. You can twist and contort and deform the balloon in any number of ways; yet, the balloon will not explode.

This level of stress is actually more desirable than the absence of stress. Interpersonal Communication specialists--Berko, Rosenfeld, and Samovar--call this desirable stress "eustress."[7] The prefix "eu-" denotes "good." We find the prefix in such terms as "eulogy" which means "good words" (spoken at a funeral) and "euphemism" which refers to a word that "sounds good" and is thus substituted for a less desirable word.

A basketball player who is "pumped up" will generally be more successful than one who is "flat." Yet that same basketball player may become so stressed that s/he misses key free throws down the stretch. This over-stressed condition would no longer be "eustress." A public speaker or actor needs to maintain a certain level of stress, or s/he may deliver the message in a lackluster manner. Yet, that public speaker or actor would not be well served to have so much stress that s/he forgets a line or exhibits nervous physical characteristics.

In everyday life, we also notice times when we are in need of eustress. If our lives are so devoid of stress that we feel bored; we are in need of eustress. If there are no conflicts in our interpersonal relationships, we humans will sometimes argue for the sake of arguing. This is a sign that our powers for dealing with stress are in need of some exercise. Similarly, although domesticated dogs are provided with dog food, there is a need for the stress of being a "hunter" which is not satiated by simply providing for their nutritional needs. Therefore, if we purchase a tug-of-war play toy, the dogs will derive almost endless satisfaction in the struggle to pull at one end of the toy while their masters pull the other end. Dogs

---

[7]Roy M. Berko, Lawrence B. Rosenfeld, and Larry A. Samovar, *Connecting: A Culture-Sensitive Approach to Interpersonal Communication Competency,* 2d ed. (Ft. Worth, TX: Harcourt Brace College Publishers, 1997), 218.

will chew and pull on rawhide bones and toss them into the air to simulate the struggles that their bodies crave.

When humans feel the need for stress, we often provide for it symbolically. We might watch our favorite athletic team play an important game. If the score is close throughout the contest, we feel the same adrenaline rush that we would feel if we were facing some significant stress in our lives. Other humans will provide the stress by watching a horror film. Vicariously, we participate with the hero/heroine and experience the same fears and stresses s/he experiences. Why do we view these stressful presentations? We view them because we actually crave a certain level of stress. This desirable stress is called eustress.

When my wife and I began our stints as students at Indiana University, we became Hoosier fans and began to watch Big Ten football and basketball contests. Our close friends were fans of the University of Illinois. When the two schools played each other, there was increased stress, as we vicariously competed with each other, through identification.[8] We had other friends who were fans of Indiana's intrastate rival, Purdue University. The stress generated by Indiana-Purdue contests was enormous. Yet, much of the enjoyment of watching Big Ten competition was the eustress that the competition produced! A certain level of stress can actually be fun!

Distress

Although some stress is enjoyable, too much stress can be experienced negatively. Consider the balloon discussed earlier. While it is quite malleable in its just-barely-inflated state, when it is inflated further, it is less accommodating to poking and prodding. Even with this medium-inflation, however, it would take strong prodding to cause it to burst. When humans, in a similar state, feel less

---

[8]Kenneth Burke, *Dramatism and Development* (Barre, MA: Clark University Press, 1972), 27-28. Burke points out, "we spontaneously identify with some groups or other . . . and . . . we spontaneously, intuitively, and often unconsciously, act upon ourselves." This spontaneous identification can be deliberately induced as when "a politician who, though rich, tells humble constituents of his humble origins." Secondly, it can operate antithetically, "as when allies who would otherwise dispute among themselves join forces against a common enemy" such as was the case in the Persian Gulf War of the early 1990's, Operation Desert Storm. But Burke claims, "The major power of 'identification' derives from situations in which it goes unnoticed." Burke's "prime example is the word 'we,' as when the statement that 'we' are at war includes under the same head soldiers who are getting killed and speculators who hope to make a killing in war stocks." In few situations does this third type of identification work more powerfully than in those situations where the terms "we" and "you" are used by fans watching athletic competitions. Effectively, rooting for opposite teams in sports or opposite political parties in an election can produce tremendous stress. Sometimes this stress is good. Sometimes, the stress is bad.

willing to "give" in to pokes and become more "rigid" in their attitudes, they are frequently experiencing heightened stress in their lives.

Taking the process of inflation to the limit, we envision a balloon that has become so stretched to contain the air pressure that it can be stretched no further. At this stage, even the lightest finger poke will produce an explosion. In human terms, we have all witnessed occasions when an unsuspecting-but-benign individual "teases" a friend in a gentle way. Suddenly, the "teased" one explodes in a vituperative fury! The "teaser" is aghast! How could such a gentle prod produce such vituperation? Simply. The "teased" one had reached the maximum stress level. S/he had been inflated to the limit. All that was necessary to produce an explosion was a tiny prod.

Clearly, having zero stress is not an option for humans. We crave some level of stress. On the other hand, having too much stress is not acceptable either. The secret to human happiness as it regards stress is to constantly maintain some medium level of inflation in the balloon. If there are times in which more stress is desirable, it is useful to know how to add stress--how to inflate the balloon by finding stressors. If there are times in which less stress is desirable, we need to be aware of the available "relief valves"--the ways in which the balloon may be deflated. Inflating the balloon is easy for the entrepreneur: go into business for yourself. Your balloon will inflate rapidly.

The Seven C's

Deflating the balloon is also a proposition that is simple to understand, but it is often difficult to achieve. If you would keep stress under control so as to avoid explosions, you must be able, first, to diagnose the cause(s) of the stress (i.e., the stressor) and, second, to avail yourself of the specific "relief valve(s)" for each stressor. The specific relief valve (correlating to the stressor) can let the pressure out of the balloon. Regardless of whether a given participant is the stressed person or the person communicating with the stressed person, it is expedient for all to be aware of the seven basic stressors and their corresponding relief valves.

As a mnemonic device (to aid in remembering the seven stressors and, thus, to facilitate easy diagnosis), I have used alliteration. Each stressor begins with the letter "C." If you are a good navigator and chart your course using the information that this book provides, you can successfully "sail the Seven C's of Stress!"

Corporal Stress

The very first place to look for stress is in the "body." I call this stress "corporal stress." "Corporal" punishment (if it is even used anymore!) is punishment of the body. A "corpse" is the dead body (after the spirit has gone from it). The Army "Corps" of Engineers is a body of soldiers who work on engineering projects. A "corporation" is a group of individuals who are treated, according to the law, as one body.

Clearly, in light of the balloon paradigm, the body has certain balloon-like organs, which can be over-inflated. The urinary bladder can be so stressed that an elementary school student, for example, can think of nothing other than utilizing the appropriate "relief valve."[9] Regardless of what the teacher might be saying to the class, the student might be groaning and waving his/her hand violently to ask permission to go to the restroom. Perhaps even the euphemism "*rest*room" is an appropriate allusion to the state of stress-relief, which such rooms exist to facilitate. In more severe cases, the intestines and colon may become distressed by over inflation producing constipation, diarrhea, or flatulence. The stomach may also become distressed, resulting in nausea. The body has obvious relief valves for such corporal stresses. Other corporal stress relief valves are not so obvious.

In Chapter 2, I provide multiple examples of corporal stress. Among other relief valves, laughter is a symbolic relief valve which, so far as we know, only humans possess. Laughter goes a long way toward relieving stress. Physical comedy is, in fact, based upon the relief that comes by seeing humor in corporal stress situations. Even tears that are often evoked by expressions of pity--sympathy or empathy--serve as relief valves for stress. What is not so obvious is that purely human symbolic relief valves such as humor and pity help to relieve even purely corporal stresses. It can be argued that animals experience essentially the same types of corporal stress as do humans. However, animals (without the assistance of humans) are limited to purely physical (non-symbolic) relief valves. Kenneth Burke offers the example of a stressed bird caught in a college classroom. All windows in the classroom were high--nearly extending to the ceiling--and open at the top. Yet, the bird kept trying to escape by flying into the ceiling rather than heading slightly lower to one of the open windows. The bird's instinct directed it to escape upwards rather than downwards toward windows.[10] Humans, on the

---

[9] Similarly, Burke, *Language*, 447, compares the cathartic release in tragedy to the "diuretic release" in bodily matters.

[10] Burke, *Language*, 3-4.

other hand, may avail themselves of instinctive physical relief valves plus symbolic relief valves such as laughter and tears plus composite relief valves. By composite relief valves, I refer to those physical remedies that are discovered by human invention (symbol-use). Specifically, I refer to the human discovery of drugs, correct nutrition, medicines, surgical procedures, *etc*--what the 17th century Dutch philosopher Spinoza calls "adequate ideas."[11]

Community Stress

Community stress is the stress that one feels when one must deal with other people. Clearly, a certain amount of this stress is desirable--hence, it can be seen as eustress. Loneliness is a word that we use to describe the condition that demands more community stress. We do need the stress which comes from having other people around us. We humans are "social" beings. Yet, because humans often disagree, stress between humans occurs. Kenneth Burke claims further that human conflict is inevitable. His philosophy is summarized in the equation, "Things move, people act."[12] Since action is his key term for understanding what humans do, he calls his philosophy Dramatism. "Drama is the culminative form of action." And if humans engage in drama, there is naturally conflict.[13] What drama have you seen that does not involve conflict and tension? Solving disagreements requires the use of community stress relief valves.

Every country has a challenge: to solve the community stress produced by the conflicts of its citizens. Some countries use the "tyrant" relief valve. Disagreements are solved by having one person (the tyrant) make the decision. Some countries use the "democracy" relief valve. Everyone votes on an issue and the course of action receiving the most votes wins. Some countries even allow a certain level of anarchy. No one rules. Everyone decides for himself or herself what to do. In solving community stress in a business or family, no single relief valve is always the best choice. The goal is to use the right valve in each instance. When it is essential that all individuals conform to a group decision, democracy or tyranny or alternating tyranny would be best. In alternating tyranny, each person takes a turn at being tyrant or each person is tyrant over one issue. For instance, I call my wife the "tyrant of taste." She chooses the color scheme and decorations of the home. She usually chooses even my clothes. She has more "taste" than I,

---

[11] Burke, *Language*, 430.
[12] Burke, *Language*, 64.
[13] Burke, *Language*, 54-55.

so she is the tyrant. When our children were young, every Tuesday night was "family night." We did things as a family but, on a rotating basis, we all took turns as tyrant deciding in which activity we would engage on a week-by-week basis.

Cash Stress

For the vast majority of human beings, cash stress is best described as the tension that is produced by a desire for spending more money than is actually available. Many businesspersons and their customers/clients feel some level of cash stress. Even when one's gross income is quite high, one may feel cash stress. There always seems to be something else to buy, some other trip to take, some new bill to pay. Admittedly, cash stress is relieved somewhat when the income and expenses are more in line but, for most of us, the cash stress is always there to some extent.

The relief valve for cash stress (for the non-rich) works in two ways. One lever that relieves stress is the one that "increases the income." The other lever is that which "decreases the expenses." Money management is the relief valve for this type of cash stress. Lest one assume, however, that the mighty increase of income would solve all cash stress, note that some of the greatest cash stress of all is the cash stress felt by the rich. The stress now relates to such issues as:

1. Is my money safe?
2. Is my money earning the highest interest?
3. What if I own stock and the market crashes?
4. What if the stock market skyrockets and I own no stock?
5. What if rapid inflation destroys the value of the dollar?
6. What if somebody sues me for malpractice, *etc.*, and takes me for everything I own?

Various relief valves for the rich might be malpractice insurance, diversification of assets, *etc.*

These are all symbolic relief valves because cash (or money) is totally symbolic.[14] Virtually anything can be converted into the money symbol. Time is money. You can put a dollar value on your time. Lodging, food, clothing, medicine, and transportation can all be translated into a monetary value. Hence, relief valves for cash stress are not usually purely physical, although some purely physical relief valves (such as sleep) work for all types of stress. The problem is

---

[14]Kenneth Burke, *Attitudes Toward History* (Berkeley, CA: University of California Press, 1984), 194.

that someone with a great deal of cash stress may not be able to sleep until s/he uses a symbolic relief valve to deal with the cash stress.

Chrono-Stress

I once visited Walt Disney World with friends on New Year's Eve. Much of the stress we faced at that time was chrono-stress. Chrono-stress is the stress one experiences when there seems to be insufficient time to accomplish one's tasks. That New Year's Eve, we simply ran out of time. It took longer than we had anticipated getting into the Alien Encounter ride. Then, it took longer than we had anticipated for some last minute shopping for a hat. Then, with the huge crowds, it took longer than we had anticipated making our way back to the exit area. The chrono-stress mounted minute by minute as we rushed in vain to reach our preferred destination to usher in the New Year.

Chrono-stress is felt by drivers who get caught in rush hour traffic jams. Many are under pressure to arrive on time, but are hindered. Shouts, honking, gestures, and (increasingly) violent responses erupt as the chrono-stress threatens to explode proverbial balloons. Sometimes, such chrono-stress is unavoidable. But it must be controlled.

My college students feel chrono-stress. They have assignments, which must be completed and turned in on time. They have exams for which they must be fully prepared. They pull "all-nighters" because of chrono-stress. Then, they may fail the exam after all! They may have so much "corporal stress" from not having any sleep that they cannot concentrate. They may blame their low grade on me, feeling that I should not have required them to take an exam under such circumstances. This attitude toward me would be a form of "community stress."

Unlike corporal stress, cash stress, and community stress, chrono-stress has only one relief valve--"time management." In cash stress, we may seek relief either by increasing income or decreasing expenses. We have no such luxury in chrono-stress. There is no way to increase the amount of time which we have available. All human beings are equal in this regard--we all have just twenty-four hours each day. Our only source of relief is in the management of this commodity.

Competence Stress

Not only do students and drivers face chrono-stress, but they also face competence stress. The first time you sat behind the wheel of a car and started the engine, your hands were clammy; your muscles were tense; you sat up stiffly in your seat with your hands at the ten o'clock and two o'clock positions; your foot was pressed firmly against the brake pedal. As you drove, you worried that you would cross the middle line of the road or run off the side of the road. This is competence stress. You weren't certain that you were competent to drive. Students feel muscle tension and clammy hands as they prepare to take an important exam or give a speech or make a presentation. They are not certain that they are competent to perform these tasks. The word "compete" is in the word "competence." Athletic performers of all varieties experience competence stress. In an important game, they may not be certain that they are competent to play a given opponent. If an individual begins a new job, s/he may not be certain that s/he is competent to perform all of the expected tasks. Let's face it. Everyone is incompetent at something. Michael Jordan may be the best basketball player in history, but he could not make it into major league baseball. Abraham Lincoln was a tremendous communicator, but he could not avoid a civil war.

We may avoid competence stress by refusing to participate in those activities in which we feel incompetent. We may decrease competence stress by thorough preparation and practice. The drivers who were so stressed out the first time behind the wheel eventually reach the point at which they sit back in the seat, set the cruise control, manipulate the wheel with one hand and eat sandwiches with the other. The students who were stressed out at their first exam may so thoroughly prepare for the second exam that it becomes fun. The basketball player who felt incompetent may practice so hard that s/he asks for the ball in the critical minutes of the game. Avoidance, preparation, and practice are among the relief valves for competence stress.

Confusion Stress

Have you ever waited until the last minute to leave for an important engagement only to find that you have misplaced your keys? Your stress explodes: "Where are my keys?" You accuse those around you. "Who took my keys? They were right here!" Your heart pounds. Your face becomes flushed. You are experiencing confusion stress. Of course, in this instance, your confusion stress is compounded by chrono-stress--you waited until the last minute--but it is confusion stress, nevertheless. Perhaps it is not your keys that you misplaced. Perhaps it is a plane ticket or an assignment or your glasses or an important bill or your diamond ring. If you can't remember where you may have left your wallet or purse, the issue is not chrono-stress, at all. The stress is confusion stress, though perhaps compounded this time by cash stress. If you lost the play tickets, your confusion stress may be compounded by community stress as you anticipate your companion's displeasure.

One relief valve for confusion stress is organization. For years, I had stacks of papers lying around my office. Whenever I needed something, I burned precious hours going through everything. I invested $29.00 in a 36-slot organizer. Even those papers that needed to be filed later had their own slot. One of the issues that always produced community stress in my family was the ice cube issue. You are right. This does sound trivial. But, every time I needed ice, I found that the ice bin was empty. I emptied a tray with 14 cubes and refilled and replaced the tray. I took four cubes for myself. Later, when I returned for more ice, the bin was again empty. "Who keeps using the ice without emptying and refilling the trays?" Of course, no one was guilty. Then, for Christmas, amazingly, Santa placed in everyone's stocking an ice cube tray, which had been personalized. I had my own ice bin and my own ice cube tray. The confusion stress (and the resultant community stress) vanished. Why didn't we invest in an automatic icemaker? At the time, cash stress.

Conscientious Stress

The seventh and final "C" which must be navigated is conscientious stress. It is that stress which humans face due to morality. According to Kenneth Burke's definition of human, humans are "the inventor[s] of the negative, (or moralized by the negative)."[15] Burke attributes the concept of the "propositional negative" to Bergson's *Creative Evolution*. For Bergson, the thought "It is not!" is something which only humans think. Dogs, for instance, do not think in terms of the negative. They do not look at Purina Dog Chow and think, "This is not Lucky Dog; this is not Kibbles and Bits." But humans can look at a knife and go on forever stating what it is not: It is not a fork, a spoon, a cow, a house, a road, a river, a star, *etc.* What Burke considers important about the human use of the negative, however, is what he calls the "hortatory negative." This use is represented by the statement, "Thou shalt not!" Being able to say, "Thou shalt not!" implies free will. It implies morality. It implies choice.

There are some things that humans cannot moralize. Suppose I said to you: For the next five seconds thou shalt not allow thy heart to beat! You would laugh. You cannot control the beating of your heart. But, if I told you: Do not steal the

---

[15] Burke, *Language*), 9ff.

five-dollar bill from my shirt pocket! you would have a choice. You would have free will. Hence, you would have morality.

There is a stress that results from this moral choice. Let's say you consider yourself to be a person who would never steal, but you walk into a restroom and discover that someone has left a wallet on the shelf with a one hundred dollar bill easily visible. No one else is in the restroom. You have an urgent bill that must be paid and you are one hundred dollars short. You think, I can either take the money or I can check the identity in the wallet and contact the person in order to return the money. You may be facing conscientious stress.

I had an Old English Sheepdog named Chaucer. I had trained Chaucer to obey my commands in virtually all areas. If I said, "Roll over," he rolled over. If I said, "Stay," he stayed. But, once there was female-dog-in-heat that lived one-fourth mile from our home. Every time we let Chaucer outside, he made a beeline for the female dog. I shouted, "Chaucer, stay!" It did not work. I spanked him. It did not matter. The female dog was kept in a barn. The barn door had a crack in it, and Chaucer had to be retrieved each time from the barn with his face bloodied from trying to squeeze through the crack in the door. He could not control himself. Amazingly, humans are able to control these drives! Moral codes may say, "Thou shalt not have sex outside the bonds of marriage." And, humans are able to conform to these moral codes!

Conscientious stress also applies in areas that are not normally considered to be moral issues. I may tell myself, "Thou shalt not eat foods high in fat or cholesterol." This is not a moral issue in most cultures. Nevertheless, I may experience enormous stress when someone places a greasy pizza in front of me. This is also conscientious stress. Whenever we are charged with a "Thou shalt not," whether or not it is part of a moral code, we are confronted with a potential for conscientious stress.

The relief valves available to us range from a refusal to obey any Thou-shalt-not's to an absolutist/legalist approach to a thoroughgoing reliance on forgiveness. If I have no moral code whatsoever, I will experience no conscientious stress. If I follow a legalistic approach thoroughly, I will experience minimum stress. I will know exactly what I may do and what I may not do. I will consider the decision to have been made in advance. Few laws are easily interpreted so strictly, however. A much more realistic approach is the committed attempt to follow our various moral codes while taking advantage of those avenues of forgiveness that various religions, states, cultures, and individuals may offer.

These are the seven C's (or sources) of stress--corporal, cash, community, chrono, competence, confusion, and conscientious. These seven C's will be addressed and applied throughout this book. Businesspersons attempt to reduce stress in their own lives, the lives of their customers, their employees, their families, and their peers and superiors. Businesspersons will discover in this book a method of reducing stress that is not a band-aid approach. They will reduce stress in the same way they design successful business proposals. They will identify the cause of the stress and develop solutions that address that cause. They will diagnose the specific stressor and utilize those relief valves that correspond to that specific stressor.

Chapter 2

Corporal Stress

Corporal stress is any stress that is encountered primarily due to the functions of the human body. The word root "corp" means "body." When someone dies all that remains is the corpse. When children are spanked, they receive "corporal" punishment. The other six types of stress are all symbolic in nature (stresses that come from our use of language), but corporal stress deals with our nature as sheer animals. Burke says that humans are "bodies that learn language."[16] For Burke, this is the same as saying the human is the "symbol-using animal."[17] In terms of the functions of our bodies, we are similar to all other animals. The difference between human stress and animal stress is that animals only experience corporal stress while humans experience the six additional stressors. Animals certainly experience disease, injuries, hunger, fatigue, thirst, the need to urinate, the need to defecate, sexual tensions, etc. These are all corporal stresses.

On the other hand, animals do not experience cash stress. Essentially, the use of money is a product of our use of language. Animals cannot even understand the

---

[16] Kenneth Burke, "Bodies That Learn Language." Lecture at University of California, San Diego, 1977.
[17] Burke, *Language*, 3.

concept of money. Animals do not experience chrono stress. I love the lyric from the musical "Big River" that exclaims, "What's time to a hog?" The calculation of time is a product of our use of language. Animals do not experience conscientious stress. They have no conscience, no morals, no ethics. These are related to the human use of language (what Burke calls the hortatory negative): "Thou shalt not."[18] Animals do not experience competence stress. They generally understand their own limitations. A dog does not sit around and worry about his inability to fly. However, a human might just do that. This is especially true if that human is taking flight instruction and she now must demonstrate to her instructor that she can take off, fly, and land safely. Her instructor has told her (using language) how this is accomplished. The only confusion stress a dog, for example, might have is trying to figure out what it must do to receive a treat from a human: Shall I beg, sit, roll over, bark, etc.? However, the stress involved here is the corporal stress of appetite/hunger. The dog would feel no confusion stress whatsoever if a treat were not at stake. So, this is not true confusion stress. Humans, however, experience confusion stress if they are unsure of what is expected from them (even if no food is involved).

Community stress is the only other type of stress that *might* be experienced by animals. When my cat, Hi-Yo--she was a "silver" cat; if you are familiar with the Lone Ranger, you will see the connection--had kittens, you might say she eventually developed some community stress. At first, there was no community stress because she instinctively became totally selfless. Everything she did was for the sake of her kittens. She scared away other animals that might have been threatening to her kittens. She spent endless hours cleaning her kittens and allowing them to suckle. Later, she killed birds and mice, but she did not devour them herself. She brought them to her kittens. After her kittens had grown a bit, she played with them; she wrestled and fought--but only for the sake of the kittens. They needed to be prepared for the corporal stresses of the real world. Once they were thoroughly trained, however, she began to exhibit what might be described as some community stress toward them. No longer was she totally unselfish. If a bowl of milk was poured, she would push her grown kittens away and devour it herself. If her offspring became persistent, she hissed and growled at the offspring. She swiped at them with her claws.

Is this community stress? Not really. After observing this changing ritual for thirteen years, I conclude that her fighting with her offspring when they had grown

---

[18] Burke discusses the human concept of morality in his "Definition of [Hu]Man" in *Language*, 9-13.

was just as instinctive as her nurturing that offspring when they were tiny. Her behavior was built-in, programmed, instinctive. It was a corporal stress--not a community stress. It originated in some bodily mechanism. Humans, on the other hand, experience community stress over different interpretations of a single word. They may even agree to share the same plate of food (as Hi-Yo would not with her grown offspring), but they disagree on what the exact nature of that food should be--pizza, burgers, salad, etc. The only stress that is clearly common to both humans and animals is corporal.

It is important to realize that, while humans do experience six symbolic (or language-related) types of stress, it is easy to confuse these other types of stress with corporal stress. Most other books on human stress that I have read do just that. They confuse the other six stresses with corporal stress and then treat the human only for corporal stress. They offer corporal stress relief valves for all types of stress. Why does this confusion occur? The answer is clear. Virtually all stress (regardless of which of the stressors is responsible for it) eventually becomes translated into corporal stress. Burke suggests there is abundant evidence of the symbolic being translated into the corporal. An anthropologist believes that the meal he has eaten is whale blubber even though it is actually dumplings. He goes outside the igloo to vomit. A tribesman enters his tent to find out that he has been "hexed." "[T]hose in authority have decreed his death by magic, and he promptly begins to waste away and die under the burden of this sheer thought."[19] Let me offer more examples.

Translated Corporal Stress
A job applicant experiences competence stress about the upcoming interview--she is not confident that she will perform well in the interview. Hence, competence stress gets translated into corporal stress: The applicant is restless, fidgeting. The applicant's mouth and throat become dry. The applicant's hands and/or feet become cold or sweaty. Her heart beats faster. She bites her nails. Her legs get shaky or tighten up. Her urinary bladder becomes more active.

She is hired, but now she is confused at her new job. No one has explained procedures. No one has trained her on the equipment. If a customer asks her a question, she probably will not have the answer. She has confusion stress--she does not know how she is expected to answer questions. She feels the same way she felt the first time she was separated from her mother at the department store

---

[19] Burke, *Language*, 7.

when she was a child. Her confusion stress is translated into corporal stress. She bites her lip. She averts her gaze from customers. She does not wish to be spoken to. Her hands fidget nervously. She paces the floor.

A cashier slips money from the cash register into his own pocket. He thinks he has found a way to keep the company from finding out what he is doing, but his conscience is bothering him. He is experiencing conscientious stress, but it is quickly translated into corporal stress. He is unable to sleep; he experiences headaches, heartburn, diarrhea. He drinks alcohol to soothe his conscience. He experiences dizziness. His stomach is constantly upset. His muscles cramp.

A commissioned sales person goes into a sales slump. The unpaid bills start piling up. The sales person experiences cash stress, but the cash stress is translated into corporal stress. Depression or aggression is exhibited in the sales presentation. The salesperson grouches at the prospective customer or he behaves in a hopeless manner: "You probably wouldn't be interested in anything like this, would you?" The salesperson taps his fingers. His neck and shoulders tighten up and ache. At night, as he sleeps, he grinds his teeth. He loses his appetite or he gains a voracious appetite. He withdraws from people and goes to bed to escape from his stress. He cannot concentrate on other tasks that need to be done. He cries privately.

A restaurant customer has been waiting for a meal that was promised to her thirty minutes ago. She is on a tight schedule. She has chrono stress. She glances repeatedly at her watch. Her face is flushed; her pupils dilate; her muscles tense; her breathing speeds up. She scoots her chair around noisily. She clears her throat. She lights up a cigarette and taps her fingers impatiently.

A college student is assigned to do teamwork with some fellow students on a project. He finds that the other team members nonchalantly dismiss every concern he raises. All decisions are made by two members of the team and he is simply given assignments. He believes the team is taking the wrong direction, but the two decision makers ignore him. He has community stress, but it becomes corporal stress. He clenches his teeth. His voice continually increases in volume. The veins in his forehead and face protrude. He cannot remain seated. He stands and pounds his fist on the table.

Originating Corporal Stress

In the above instances, I do not recommend treating the stress as corporal stress. What became corporal stress originated from one of the other six stressors. I recommend diagnosing the originating cause of the stress and providing a relief valve appropriate to the originating stressor. The remainder of this chapter, however, will emphasize the stress encountered PRIMARILY due to the functions of the human body, the animal nature. I refer to any bodily stress that exists independently of any of the other six stressors. Although a thorough discussion of such corporal stress and the appropriate relief valves would require a medical library--not just a chapter in this book--this chapter can at least point to the broad types of corporal stress that may affect you, your friends, and your business associates. If you find that your stress is primarily corporal in nature, consulting a physician is one most obvious solution. The physician should be able to prescribe specific relief valves for your specific corporal stress.

An example of my point here is the treasurer of an organization with whom I was associated. He was always a very easy-going person. Nothing seemed to fluster him. He looked at life positively. He wrote jokes for the organization's newsletter. If funds were slow in coming in, he just notified creditors that the bills they had sent might be paid a few days late. He worked with people. I cannot remember him even displaying a temper, until . . . one year. The organization was thriving. Finances were excellent; there was even a considerable surplus. Gradually, he began to complain about little things. He now believed that employees who had not received a raise in two years were making too much money. Despite the increasing financial strength of the organization, he worried aloud that the day might arrive when the organization would not have the funds to pay the bills. Drastic changes needed to be made. One day he recommended cutting the payroll in half. The next day he recommended doubling the payroll. He wrote scathing notes to his peers and superiors.

What happened to this mild mannered Clark Kent type? His physicians diagnosed it--cancer. Cancer produces not only corporal stress but also confusion stress. The patient is lost, confused. He does not know whether he will live or die. This means that he probably also has chrono stress. He may not have enough remaining time to accomplish what he wants to accomplish. He may also have cash stress. Medical bills may not be fully covered by insurance. Lost time at work may reduce the paycheck. He may wonder if he has enough life insurance to care for his family. The balloon may reach extreme distress levels. Even the slightest prod may cause it to explode.

I once provided some free financial consulting for a friend who found out he had cancer. I did not recommend meditation or deep breathing exercises to relieve his stress. I certainly could not cure his illness. Instead I left that to his physicians and tried to help relieve some of his other stresses. I examined his financial portfolio. I advised him to convert his company term life insurance to permanent coverage before he quit working. I recommended that he exercise a guaranteed insurability option he had on one of his personal life insurance policies. I helped him get all of the life insurance benefits that were possible for him. Next, I discouraged his wife from planning to work full-time when he died. The couple had several young children and she would receive full social security benefits for herself and her children if she did not work full-time. My friend lived his remaining days more peacefully because I had helped to relieve some of his cash stress and chrono stress, even though I could do nothing about his corporal stress. The physicians relieved some of that stress for him--through medication. As indicated, corporal stress often carries with it other types of stress. Frequently, you can assist in the relief of these other stressors, but if you sense that a friend or business associate is stressed out, the place to begin your diagnosis is with corporal stress. If the stress is originating in the body, you typically must refer the friend or associate to a physician. However, there are some things you can do to reduce corporal stress in the home and workplace.

Relief Valves for Corporal Stress
Taking Protective Measures. Using common sense and taking protective measures is a powerful relief valve for corporal stress. Those who are involved in automobile accidents generally experience less corporal stress if they have taken the protective measure of wearing a seatbelt. Football players experience less corporal stress if they wear helmets and padding. Protective measures are effective in reducing corporal stress. Not every disease bug can be avoided but some can be. The old saying, "Cleanliness is next to Godliness," may apply in the arena of corporal stress.

A hospital neonatal care unit in the South experienced an extremely disturbing problem. Babies in the unit were dying. Statistically, newborns do have a higher mortality rate than one-year-olds, but in this unit the rate of mortality for the infants was far beyond levels in other hospitals. What was causing the problem? A study found the culprit. Nurses in the unit wore fake fingernails. These long nails provided a breeding ground for disease. According to the *Detroit Free Press* (October 5, 1999), fake fingernails are deceptively dangerous. While they appear to be clean, "they are home to a host of potential pathogens, including yeasts." One study found that, while one-third of nurses with natural nails had pathogens on their hands, the number of those with pathogens on their hands jumped to three fourths among nurses with acrylic nails. The study recommended that nurses avoid fake nails or at least wear rubber gloves. If the lives of newborn babies can be threatened by what one would consider the clean hands of health care workers, everyone is at risk of corporal stress from the transmission of disease through hand contact. Sparklingly clean and disinfected restrooms in the home and the office are certainly helpful. Frequently disinfecting door handles, faucets, flush handles, locks, and railings should reduce disease transmission. Frequent and proper hand washing, consciousness of the diseases passed by handling paper currency, and being on the lookout for potentially accident-producing circumstances in the home and office reduce other types of corporal stress.

Besides reducing stress, it is pragmatic for businesses and families to emphasize safety and cleanliness. McDonalds found it sold more hamburgers than other establishments by advertising, "At McDonalds, it's clean!" How much more would you prefer to frequent a clean business than a dirty one? Then, why not view your own home as a business? As McDonalds' jingle once said, "You deserve a break today!" Give yourself and your family a break from corporal stress.

Obviously, we prefer cleanliness in our restaurants, but even businesses not typically considered disease-related could benefit from cleanliness campaigns. I found a radio commercial amusing. The ad was promoting a plumbing service. The company promised promptness in responding to calls--something you might think would be important in their business. Then, the ad also promised something that took me by surprise: it promised that the plumber who came to the customer's home would be clean! Without saying anything negative, the implication was clear. Other plumbers who come to your home may stink, but ours do not. I guess, regardless of the business, cleanliness sells! And, even if the corporal stress is only the stress our bodies experience when our noses encounter a disagreeable odor, that disagreeable odor is a corporal stress, nonetheless! Have you ever left a place of business because it smelled funny? I know I have. One grocery store in my hometown lost my business because of smell. My guess is that, if a grocery store smells bad, I might also risk corporal stress by eating food from the establishment. In the home, smell is often an indicator of disease risks, air-borne fungi, or allergens. Any of these might produce corporal stress.

Clearly, everyone at times enters environments that might pose risk for corporal stress. Even if you wash your hands frequently, you may encounter germs on the door handle as you exit the lavatory. Consider keeping a supply of alcohol-based wiping cloths at your desk, in your car, wherever it is convenient. Additional protective measures might include flu shots and pneumonia shots. You might also look for a health insurance plan that pays for preventative health care, with no deductible.

Considering all the Bodily Systems. Despite a strong preventative plan, some illnesses will occur. A wise employer will recognize that all systems of the body are subject to corporal stress. I know something about these bodily systems. At Indiana University Kokomo, I taught the course, Greek and Latin Elements in Medical Terminology. My doctorate is a Ph.D., not an M.D. I do not claim to be a medical expert. This course was a Classics course. I do have some expertise in classical languages. However, teaching the course forced me to become familiar with many of the stresses of the bodily systems.

The **cardiovascular system**, for example, has internal pressures (or corporal stresses). The absence of a pulse or heartbeat indicates a virtual absence of stress. This is not a good situation. We want the stress of a pulse. Low blood pressure is an example of a bodily condition that cries out for more stress--eustress. Hypertension, however, indicates that the blood vessels are receiving too much stress. High cholesterol contributes to the stress that is placed on the veins and arteries. Medication, surgery, exercise, and diet are composite relief valves. Soothing music, human support and reassurance, and humor are symbolic relief valves. Sleep or relaxation, with the accompanying reduction of pulse rate, is a purely physical relief valve.

The **respiratory system** has its own forms of corporal stress--asthma, bronchitis, pneumonia, and emphysema. As mentioned earlier, you may want to avoid locations that smell bad or give indications of disease risks, air-borne fungi, or allergens. Allergies run in my family. My brother had hay fever every year. I was allergic to cow's milk as a child and my skin was very sensitive to plants of several varieties. When my son began to exhibit respiratory corporal stress, we took him to the physician to undergo allergy tests. Among other things, he was allergic to pet dander. The cats were moved outside. We changed our furnace filter to the variety that filters out air-borne allergens. These things seemed to work as relief valves. Flu shots, asthma medication, antibiotics, avoidance of smoking, avoidance of allergens, oxygen masks, mechanical respirators, *etc.*, are composite relief valves. Soothing music and talk may be symbolic relief valves even for this corporal stress. Sometimes, your body naturally fights the corporal stress, by itself. Natural bodily defense systems are purely physical relief valves.

The **nervous system** may be plagued by a pinched nerve, a herniated disc, or even a tumor. Of course, tumors can cause corporal stress to virtually any system of the body. I have several friends with back problems. In some cases, the problem is due to athletic injuries. Many parents are inclined to encourage the athletic participation of their children because American culture values athletic prowess very highly. Yet, coaches and parents who consider risking injury to their children's nervous system for the sake of some fleeting athletic honor should think again. Frequently, the damage to the nervous system cannot be reversed. The only effective relief valve for some injuries to the nervous system may be to avoid the dangerous activity initially. Even so, traction and surgery are composite relief valves that often work. The body's tendency to contort and adjust its position until it relieves the pain itself is a purely physical relief valve. By all means, amateur chiropractors and those professionals who are prone to produce even more injury to the nervous system than they cure should be avoided.

The **reproductive system** carries with it not only stresses caused by diseases but also stresses produced by menstrual cycles and stresses which result from the sex drives themselves. Examples of reproductive system stress abound. Many rapes and molestations are the result of male reproductive stress out of control. PMS often produces such corporal stress that the stressed individual nears the explosion point. Often, medication, surgery, and marital intimacy are composite relief valves for varieties of reproductive stress. Monogamy (combined with choosing a non-promiscuous spouse with no history of STDs) is a major preventative relief valve for avoiding reproductive diseases. Sometimes the simple recognition that you are facing a temporary, if unavoidable, corporal stress in this system will demand that you stay home from work for a few days.[20]

The **musculoskeletal system** can produce a certain amount of corporal stress. The year I wrote my doctoral dissertation, I developed an excruciating tendonitis in both of my wrists. My hand specialist blamed the problem on my incessant computer keyboarding for the year. I suspect that the condition was exacerbated by my use of a weed whip for cleaning brush out of a wooded area at my church. Regardless of the cause, the pain was so intense I was unable to swing a bat, throw a ball, play tennis, or even sign my name or use table utensils. My physician injected cortisone into my wrists, but the cure was worse than the illness. Eventually, my body healed itself with the help of wrist braces worn at night. I know that others have not been so fortunate. Carpal tunnel syndrome frequently requires surgery to correct. Businesses that require a great deal of keyboarding

---

[20] W. David Hager, M.D. and Linda Carruth Hager's 1995 book, *Stress and the Woman's Body*, is an approach to dealing with female corporal stress.

should be aware of such dangers. Hand and wrist exercises, frequent breaks, and better-designed keyboards may be preventative relief valves.

The **integumentary system** (skin, hair, etc.) can be afflicted with corporal stress. Have you ever bumped someone with sunburn? Did that person explode at you? Yes? That is proof of corporal stress produced by the integumentary system. Obviously, cuts, burns, rashes, insect bites, infected hair follicles, and skin allergies produce corporal stress. Even infants explode with dissatisfaction at integumentary stress. If their skin is in a state of distress, they scream.

The **endocrine system** produces hormones. Two of those hormones, estrogen and progesterone, skyrocket to extremely high levels in women during pregnancy and then drop severely to pre-pregnancy levels soon after delivery. According to Raphael Good, an Obstetrics-Gynecologist and psychiatrist, childbirth produces "more pronounced changes than at any time other than death."[21] A mother in Texas suffered from postpartum depression caused by this hormonal drop. As a result of this corporal stress, she felt she was a poor mother to her five children. She reportedly was given some anti-depressant medication for her problem. Whether as a result of the medication, the postpartum depression, or a combination of the two, one day the mother drowned all five of her children. This is an extreme example of distress--the woman's stress balloon exploded. But, the primary stressor was corporal (whether induced by her endocrine system or by drugs).

Speaking of drugs, perhaps the single most effective relief valve for corporal stress that can be enacted in the business world and in family life is the elimination of the use of alcoholic beverages. The power of this single drug to produce corporal stress in America is astounding. Sixty-six percent of all automobile fatalities are alcohol-related. Consider how much corporal stress is produced each day in automobile accidents. What if that stress could be reduced by two-thirds! Furthermore, most rapes and family abuse cases are alcohol-related. Fights erupt even among friends who have been consuming alcohol. A very mild example of the corporal stress produced by alcohol occurred at an insurance-company-training week in Dallas, Texas. This insurance corporation refused to even give coverage to twenty-one year old males who had been convicted of a DUI offense. The company realized how deadly alcohol was. Yet, at virtually every company meeting, the liquor flowed. I was always amazed that the company was promoting the very substance that cost the company money as death benefits were continually

---

[21] Susan H. Greenberg and Karen Springen, "The Baby Blues and Beyond," *Newsweek* (July 2, 2001), 26.

paid out. Nevertheless, one evening during the company-training week, the company provided a keg of beer. We had been conducting a ping-pong tournament all week long and that evening I was scheduled to play in the finals against a man who had beaten me soundly in every previous match we had played. The only difference this evening was that I am a non-drinker and he had consumed one glass of beer. The match began and, to my surprise, I was winning every volley. I finally led the man by a score of twenty to three. I was one point from winning the week's championship. He exploded and slammed his paddle on the table. Unintentionally, he had smashed the ping-pong ball. His explosion seemed due largely to his alcohol-induced corporal stress. We went to the cabinet for a replacement ping-pong ball, but the smashed one was the last in the supply. We would have to finish the match the next day. That next day, after the effects of the alcohol had worn off, he came back from the earlier deficit to defeat me twenty-two to twenty. What a difference one glass of alcohol makes to someone's corporal stress level! Of course, some drugs may actually reduce corporal stress. But, recreational drugs such as alcohol are well known to produce much more corporal stress than they relieve.

Beyond the systems of the body, even the eye and the ear have their own corporal stresses--pink eye, ear infections, retinitis, vertigo (dizziness), sties, ringing in the ears, and headaches. Before we assume that our stress is derived from external sources, it is important to consider any corporal stresses we may find ourselves subject to. It is always best to reduce the level of corporal stress before attacking the other sources of stress.

Universal Relief Valves

While some cases of corporal stress can be quite severe, not every case is. One symbolic relief valve for stress--which is not available to animals (except the hyena!)--is laughter. In cases where it is clear that the corporal stress is minor and temporary, laughter can be a great relief valve. Although laughter is a universal relief valve that helps us deal with a wide variety of stressors, it certainly is not appropriate for all corporal stress--let alone every other type of stress. However, for mild cases of corporal stress, laughter frequently works. Check out the get well cards in the Hallmark store. How many of them are light-hearted and humorous? Laughter alleviates some corporal stress.

Two business associates were such close friends they decided to vacation together with their families at Walt Disney World in Florida. All eleven of them--parents and children--crowded into a seven-passenger van for the drive from the upper Midwest to Florida. The physical crowding in the van was a corporal stress. When they stayed at a motel room, the eleven shared a single room. Bodies on beds and in sleeping bags covered every square inch of space, supplying further corporal stress. To make matters worse, virtually all of the vacationers came down with the flu during the trip. The motion of vomiting is actually a corporal stress relief valve. The pressures of nausea and an upset stomach are relieved (automatically and often involuntarily) by vomiting. Bodies, even of animals, have such built-in relief valves to handle stress. Such relief valves available to animals (without human assistance) are purely physical. And, that physical relief valve was in evidence those nights. Throughout the night, as one after another felt the urge to regurgitate, each found it necessary to find his/her way to the bathroom in the dark through the maze of bodies sleeping on the floor. Sometimes steps taken resulted in pain (corporal stress) for the sleeper who had been stepped on. One morning, a child fought through the maze to the bathroom and emptied the contents of his stomach in the direction of the stool . . . only to find another child sitting on the stool. The result was unpleasant, but the mother of the offended child chuckled lightly as she helped her child into the shower to wash off.

The families endured cold, sandblasting wind, and freezing gulf waters on their trip to the beach. Arriving at the beach, they shed their outer clothing, down to their swimsuits, and were hit by a cold 45-degree wind blowing in from the Gulf. The beach was virtually deserted, except for some sea gulls and a few other diehards who were determined to enjoy the beach and sunshine despite the cold wind. The other diehards, however, were wearing slacks and sweaters and, no doubt, enjoyed cracking jokes about these eleven who were running around half-naked in the chilly weather. "Just wait until we get to the water!" someone insisted. "It's much warmer in the surf." Dutifully, the others followed, fighting back pangs of doubt. At last, they plunged their feet into the warm gulf waters and, one second later, pulled out ten ice cubes per person--attached to their feet where toes used to be! Those promised warm gulf waters had been replaced by the salty brine of an ice factory!

They tried, against all odds, to salvage something from their day at the beach, but to no avail. They thought that, if they lay out on the beach to catch some rays, they would be less of a target for the chilling wind; but, upon lying down, they were sandblasted by the vicious wind. They sought refuge from the wind next to the pillars that supported a pier that led out into the gulf. This provided the best solution, thus far. It broke the wind and, consequently, hampered the sandblasting. If it hadn't been for the jackhammers operating on the pier while they were renovating the boardwalk above this group, their beach party would have had at least one redeeming feature. Could it be that they were smiling as they trudged back to the van?

They decided to have a picnic on the beach at Daytona before visiting Saint Augustine with its old Fort and other attractions. They set up a portable charcoal grill on the beach. It stood three inches tall. The wind was blowing so strongly that it was difficult to get the fire started. To reduce the wind factor, they drove the van onto the beach and placed the grill behind a tire for protection. At last, they could start the fire and prepare the hamburger patties. They hungered for the taste of freshly grilled sandwiches! Minutes later, their appetites were sated on "real SAND-wiches!" The ocean breezes had whipped up sand from the beach and had thoroughly coated their burgers to create honest-to-goodness SAND-wiches!

The cool Florida weather had not afforded them the opportunity to gradually tan their bodies throughout the week, so they threw away the suntan lotion the final day in Florida and exposed their unprotected skin to the sun's rays in one last desperate attempt to get a tan. It worked . . . a little too well! On the ride north that night the driver's body began shaking uncontrollably. His teeth chattered and he could barely talk to the motel receptionist as they stopped to rest for the evening. They then crowded one final time into another motel room--this time with the added adventure of trying not to touch each other due to sunburn pain! How did the families handle the corporal stress? They laughed.

Their trip to Florida resembled the Vacation movies starring Chevy Chase. The Griswolds had nothing on these two families. Actually, the basis of physical comedy is typically corporal stress. Humans laugh at the physical stresses of such classic comedians as Charlie Chaplin, the Three Stooges, Johnny Carson, and Jerry Lewis. As Johnny Carson cuts off the necktie of Ed McMahon and Ed retaliates by pouring some disgusting liquid on Johnny, the audience roars. The Three Stooges physically beat upon each other for our enjoyment. Jerry Lewis, Charlie Chaplin, and Peter Sellers (as the bumbling French detective Inspector Clouseau) encounter countless physical accidents--exploding bombs, dog attacks, falls through collapsing floors, running into doors and trees--all so that we may laugh at them. Laughter relieves stress. Laughing at our corporal stress and others' corporal stress can relieve our own corporal, chrono, community, competence, confusion, cash, and conscientious stress. Frequently, public speakers relieve their own stress and the stress of their audiences by beginning their speeches with a humorous story. Often, the humor is self-deprecating.

A second universal relief valve is weeping. Both laughter and tears are mild explosions of sorts. You can almost see the stress balloon swiftly deflating as each of these two universal relief valves is experienced. Not surprisingly, then, laughter and tears form the basis for our two major types of theatrical presentation--tragedy and comedy. Picture the two masques associated with the theater. One is laughing; the other is crying. Tears can relieve all types of stress as well.[22]

My daughter had just graduated from college. She had studied Music Education and wanted to teach music in a public school setting. She had no money and was cash stressed at having to pay for student loans. She did not even know if she would be able to find a job in her field. She was lost. Confusion stress. She was afraid that if she did not find a job within two months, she would need to wait an entire year to find a job in her field. Chrono stress. Every day two or three friends would ask her if she had found a job yet. She found this annoying. Community stress. Since she had not yet been hired, she began to doubt her own ability to do the required work in her field or even to be a successful interviewee. Competence stress. She even felt guilty for depending financially on her parents. Conscientious stress. She said to me, "Dad, I just feel like crying." I said, "Why don't we try that!" I hugged her and she cried, and the stress was relieved. People who are too proud to cry fail to avail themselves of one powerful relief valve. I have even seen individuals who have lost family members who refuse to cry due to some cultural display rules such as "Men shouldn't cry." I suspect that one reason men do not like to see movies that are labeled "chick flicks" is that such movies are often quite sentimental. They tend to make the audience cry and men do not want to be seen crying in public. So, they force emotions inside, producing ever-increasing stress. Why not have the best of both worlds? Rent the sentimental movies on video or DVD and exercise your universal relief valve in the privacy of your own home.

Conclusion

I was always amused that in my graduate student years at Purdue University, when taking courses under my fellow Burkean, Don Burks, we would pause from

---

[22] Burke, *Language*, 94, offers the following formula to show how tragedy relieves stress (or supplies catharsis): "Take some pervasive unresolved tension typical of a given social order [community stress, in the terminology of this book]. . . . [R]educe it to terms of personal conflict (conflict between friends, or members of the same family). Feature some prominent figure who . . . carries this conflict to excess. . . . So arrange the plot that . . . his excesses lead necessarily to his downfall. Finally, suggest that his misfortune will be followed by a promise of general peace."

time to time in lengthy class sessions for what Burks called "motion breaks." Burks was alluding to Kenneth Burke's distinction between "action" and "motion." For Burke, humans engaged in "action" whenever they used symbols/language. But, the sheer functions of the body did not represent action, because they were not the result of the human "free will." They were the result of the human's animality. Therefore, they were motions not actions. Don Burks was suggesting during these motion breaks that our bodies needed to relieve those bodily stresses that had built up in the previous hour or so. He was right. Our bodies had corporal stress. Appropriate relief valves were needed.

Chapter 3

Community Stress

Community stress is the stress one feels due to conflict with other humans. Hermits avoid this stress by staying away from other humans but the rest of us must deal with this stress. Our stress results from the fact that no two humans agree on every issue 100% of the time. Even the most devoted and loving husband and wife will at times encounter community stress with each other. They will disagree--sometimes quite strongly. Religions--even those that preach love-- will often find themselves opposed to other religions. Burke observes: "[R]eligions are so often built *antithetically* to other persuasions . . . Christianity in opposition to paganism, . . . Protestant offshoots in opposition to Catholicism, and . . . churchgoing . . . in opposition to communism."[23] Certainly, then, friendships will be tested by community stress. Employer-employee relationships will experience stress. Salespersons will have stress between themselves and their customers or prospects. There will be internal stress--stress within an organization. There will be external stress--that which exists between members of an organization or business and those outside the organization or business. There

---

[23] Burke, *Language*, 12.

will be stress between business partners, schoolmates, colleagues, and family members. Since Burke claims that humans act (and acts are Dramatistic), he concludes: "[Y]ou can't make a drama without the use of some situation marked by *conflict*."[24]   And, as mentioned earlier, some level of community stress is actually desirable: eustress.   What fun would there be in watching athletic contests, for instance, if fans did not express differing loyalties?

There is even a certain amount of enjoyment to be derived from political arguments. My best friend and I are from different political backgrounds. My parents raised me to be a Barry Goldwater/Ronald Reagan conservative Republican. The only times I can remember my folks considering voting for a Democrat were in local politics and in those rare occasions where the Democratic candidate appeared to be a conservative while his Republican opponent was clearly a liberal! It rubbed off! For the sake of certain audiences (which clearly contain strong Democrats) I have tried to portray myself at times as being truly open-minded and rather independent.   But, to be honest, the Republican chromosome is in my genes! I remember being tempted to vote for Jimmy Carter when he ran against Gerald Ford because I felt that Carter clearly shared more of my bedrock values than Ford. Perhaps, if Ford were a liberal, I would have voted Democratic, but that Republican chromosome was too dominant. When we made a trip to Florida, the spring after Carter was elected, we visited Plains, Georgia, along the way. We even attended church services at the Plains Baptist Church--Jimmy's home church--and picked up a pop can (our scruples would never have allowed a "beer can") from Billy Carter's gas station. I even felt sorry for Carter that world events, and chaos in his own party rendered him (according to many) an ineffective President in some areas.

But, when he ran for reelection against Ronald Reagan, the pride of conservative Republicans, the decision was easy. Reagan had stopped at Purdue University in West Lafayette in that campaign and I had dragged my family out to the campus in hopes of catching a glimpse of the man. We had arrived early enough to take a front-row position, waiting for him to arrive just outside the Purdue Memorial Union building. Not only did we see him, we shook hands with him. He shook hands with all of us--except my daughter Auburn. She was just two years old at the time, and somewhat xenophobic. He even stood and talked to her for a minute to coax her to vote for him. I told her, "Auburn, this is going to be the next

---

[24] Burke, *Language*, 29.

President of the United States!" Reagan smiled and concluded, "She couldn't care less!"

OF COURSE, I stayed up all night, on Election Night to see Reagan's landslide victory! OF COURSE, I celebrated! OF COURSE, I brainwashed all of my children, to make them Republicans! Somehow, through the years, I guess the chromosome affected my wife as well. Maybe, mystically, when "two become one flesh," they also become one politically. My wife's family, however, was comprised of many staunch Democrats. Perhaps it was my relationship with her family that paved the way for my relationship with my friend. Just as my folks could never vote Democratic, my wife's family could never vote Republican. Indiana, as a whole, is largely a Republican state, and usually has Republicans in most important state offices. But from my wife's family, I have never heard of anything they have ever done right! Yet, I have always enjoyed a great relationship with my in-laws. I learned to change the subject when the stress became too apparent. Amazingly, I have found, most political issues can be tabled from most conversations without great loss. My maternal grandfather, Grandpa Larson, once predicted that some day I would become President of the United States! As insightful as I know this makes my Grandpa appear to be, I have this strange feeling that if I were ever to run for the office, and the final deciding votes came down to my in-laws, and I were running as a Republican, I would be well-advised to begin preparing my concession speech.

Such was the political passion of my Democrat friend. Through the years, I have endured remarks about Ronald "RAY-GUN" and jokes about Dan Quayle. Both my friend and my in-laws have seldom let me forget the problems of Richard Nixon, Spiro Agnew, Iran-Contra, Oliver North, and others. It was humorous to watch my friend, however, when he and his wife were entertaining his wife's parents. Her parents were devout Republicans and I just sat there quietly and smiled as they razzed my friend constantly about Republican superiority. I guess it may just be that prerogative of in-laws to razz each other concerning politics.

Not only did my friend and I have the community stress of politics, but my friend and his wife are also the friends I referred to in Chapter 1 who were fans of the University of Illinois when we were fans of Indiana University. The strange thing about that particular community stress is that I had completed all of the coursework for a Ph.D. at Illinois, but I preferred to be a fan of Indiana. Why did I bring on such community stress? Why didn't I just become a fan of Illinois when I lived in Illinois? I don't know. But a similar scenario occurred when I lived in West Lafayette, Indiana (the home of Purdue University). I had managed an insurance office that specialized in working with graduates of Purdue University. I had taught courses for Purdue University for more than a decade. I completed my Ph.D. at Purdue. Why didn't I just become a fan of Purdue when I lived in West Lafayette?

I guess I should just face it--I enjoy a certain amount of community stress. I like to root for a team other than the local team.[25] As a boy growing up midway between Saint Louis and Chicago, I chose not to be a fan of the White Sox, Cardinals, or Cubs. I was a San Francisco Giants fan. I loved Willie Mays, Orlando Cepeda, and Juan Marichal. I got Willie McCovey's autograph on my scorecard the year he was rookie of the year. Even when I taught at Loyola University Chicago, not far from Wrigley Field, I did not become a Cubs fan. There is just something enjoyable about rooting for a team no one else close to you is rooting for. But, can this stress get out of hand? Can it go too far? Can it become not eustress but distress? Of course, it can. The secret of successfully sailing the seven Cs of stress is not removing all stress. The secret is controlling that stress, recognizing when the balloon is becoming over-inflated, keeping the stress at eustress levels. How can this be accomplished? It is important to understand the two sources of community stress--culture and psychology.

Psychological Stress
Psychology typically refers to the behaviors, values, and characteristics of an individual. Culture refers to those behaviors, values, and characteristics of a distinct group of people. It is clear that a psychologically normal, well-adjusted individual does not face community stress when communicating only within himself/herself. When one individual communicates with himself/herself, it is called intrapersonal communication. We carry on such intrapersonal communication, for example, whenever we have decisions to make. When I was single, I decided where I would eat most of my meals. When I dated Linda, the woman who would later become my wife, it was necessary to take her view into account. On our very first date, I discovered that she didn't like pepperoni pizza! This was stressful! Now, I had to compromise with regard to my eating preferences. When our kids were small, Linda and I made the decisions concerning meals. As the children grew, they each became increasingly assertive.

---

[25] I should point out that there are risks involved in rooting for opponents of local favorites. Burke, *Language*, 51, speaks of a "particular kind of scapegoat . . . a device that unifies all those who share the same enemy." It is easy to become a scapegoat in such circumstances.

Try to imagine the community stress that we encountered at mealtimes, as the (fully-grown and thus assertive) Lindsays had to decide together where to eat!

One relief valve for community stress is the "tyranny" valve. Even entire nations have sometimes found that having a "tyrant" make all of the decisions relieves some community stress. Decisions are no longer an issue. It does not matter what our individual preferences may be. The only question is, "What does the tyrant say?" Of course, we sacrifice some free will in the process, but the community stress may be reduced. Just as nations tire of tyrants, so individuals long for such solutions as democracy and anarchy.

The relief valve of democracy is often characterized by the four-letter word: VOTE. Sometimes this works. My family sometimes votes to eat at a roast beef restaurant. Since I do not like roast beef, I resist this suggestion. But I frequently acquiesce. I just don't eat the beef. But what if the majority prefers to eat at a very expensive restaurant? I used to typically pick up the tab at all family meals. What if I cannot afford the choice of the majority? Is voting in this instance a successful valve for the relief of community stress? No. If anything, voting is the culprit! Questions (and, of course, stresses) arise over who gets to vote? And, how much does each vote count? Should the vote of a two-year-old be equal to the vote of a mother? Should the vote of a dependent be equal to the vote of a wage earner? And what if there is a tie? Who breaks the tie? This is even a problem in marriages. What if the husband wants one thing and the wife another? Who breaks the tie?

Another democratic relief valve (that is in many respects superior to the four-letter word VOTE) is another four-letter word, which, incidentally, contains exactly the same letters, in a different sequence. I refer to the four-letter word VETO. It works many times when voting does not. If person one wants to eat at Chili's, but person two does not, person two may veto Chili's. If person three wants to eat at the Olive Garden, but person four does not, person four may veto the Olive Garden. The frequent result is that no restaurant has remained un-vetoed. In this instance, no one eats. Dieting may be better for us, anyway. Eventually, though, the corporal stress of hunger overpowers the community stress of preference and a compromise is reached. This is how the veto operates. In marriage, the husband may want to spend the tax refund on a new television. The wife may want to spend the tax refund on a new refrigerator. They may therefore veto each other's proposal. In this instance, the money is not spent. It is put into savings. Eventually, compromise is reached.

If tyranny sometimes works and democracy sometimes works, anarchy also works at times. In simplest terms, anarchy means, "nobody rules." There was a nineteenth century motto promoting church unity, that suggested: "In essentials, unity. In opinions, liberty. In all things, love." The second element of that catch phrase is a principle of anarchy. There may be instances in which each individual should have the latitude to decide for him or herself. When there is no compelling reason for everyone in the group to be doing the same thing, why not provide liberty/anarchy? If the meal were to be consumed at the mall, what would be wrong with each individual choosing a different eating establishment? Perhaps, there is even a central court where everyone could sit together. If there are no time constraints (no chrono-stress, in my terminology), why not "drive-through" several fast-food restaurants?

Cultural Stress

In psychological stress, any difference in matter of taste on the part of given individuals may provoke stress. It is somewhat easier to anticipate the possible differences when the source of the stress is culturally based. The most easily recognized source of community stress is cultural stress. If you deal with individuals from other cultures you will experience some community stress. This is due to the fact that the different cultures provided you and other individuals with different behaviors, views, and attitudes on a wide variety of issues and situations. As you begin to communicate, you will notice that an individual from another culture will not agree with you on matters you took for granted. It will seem absurd to you that anyone would even question your positions. You simply assumed that everyone agreed with your position. That position had been provided to you by your culture. A different position had been provided to the other individual by that individual's culture. This cultural disagreement produces stress. So, what is a culture? What do cultures do? And how do cultures form?

Communication specialists, Dan O'Hair, Gustav W. Friedrich, John M. Wiemann, and Mary O. Wiemann, offer a definition of culture that serves as a good skeleton upon which to build your understanding of culture and cultural stress: "We use the term **culture** to refer to the shared beliefs, values and practices of a group of people."[26] I disagree with others who stipulate that cultures must be

---

[26] Dan O'Hair, Gustav W. Friedrich, John M. Wiemann, and Mary O. Wiemann, *Competent Communication*, 2nd ed. (New York: St. Martin's Press, 1997), 9.

"a large number of people."[27] In my opinion a culture can be as small as a single family unit or two people stranded on a desert island. Those small units will still develop their own "shared beliefs, values, and practices" within their own small culture. I also disagree with those who stipulate that cultures must be developed "in the course of generations."[28] An occupational culture, for example, can develop very quickly.

I might add to the definition of culture offered by O'Hair, et. al. that culture is usually understood implicitly.[29] If I walk up to a business associate and stretch out my open right hand to him, it is not necessary for me to explain to him what he should do. He reaches out and grasps my right hand in his right hand. His grip is not too limp--firm but not too firm. He shakes my hand for approximately one second and lets go. This action may be accompanied by some small verbal greeting such as, "How are you?" How did he know to do all of that? His culture taught him. I simply assumed that he would respond in precisely the manner he did. These cultural behaviors are implicit, not explicit.

Notice also that my business associate's implicit behaviors were a set--he grasps, he grips, he shakes, he releases, and he verbally greets. What if my associate does not know my culture? What if he does not grasp my hand when extended? I feel community stress: Does he not consider me a friend? Is he snubbing me? What if his grip is too firm? I may have both community stress and corporal stress: Is he trying to harm me or seem superior to me? What if his grip is too limp? I may think: This weak individual is not my equal. What if he continues to grasp my hand longer than one second? I may have community stress: What message is he sending me? I may feel I need to pull away from him.

Practices. The definition supplied by O'Hair et. al. is pregnant with explanatory power. It gives three classifications for sets of cultural implications or assumptions. One classification is practices. Greeting fits in this classification. All humans greet one another, but your culture teaches you how this should be done. Obviously, in American male business culture the greeting occurs as a handshake in the manner I have described. But, American male business culture is only one culture. In American female culture, individuals greet each other, too. But, outside of business contexts, the handshake is not as common among females

---

[27] Robert E. Smith, *Principles of Human Communication*, 4th ed. (Dubuque, IA: Kendall/Hunt, 1995), 34.

[28] Larry A. Samovar and Richard Porter, *Communication between Cultures* (Belmont, CA: Wadsworth, 1991), 51.

[29] Hence, culture supplies entelechies. For a discussion of this matter, consult Stan A. Lindsay, *Implicit Rhetoric: Kenneth Burke's Extension of Aristotle's Concept of Entelechy* (Lanham, MD: University Press of America, 1998).

as it is among males. The hug is more common in American female culture, but there is a set of implications about how that hug will be practiced. In some cultures, the appropriate greeting is a kiss. But you must know the implicit set. It may be a kiss on each cheek; it may be a series of three kisses (from cheek to cheek to cheek); it may be a kiss on the hand, a kiss on the forehead, a kiss on the neck, a kiss on the lips. In some cultures, the appropriate greeting is a bow, but know the set. The bow may be accompanied by eye contact or by averting the gaze. It may be accompanied by folding your hands together in front of your neck and chest or by a circling gesture of one hand. The bow may be at the neck or at the waist or even with a knee to the ground (kneeling). Cultural expectations are such that the violation of one element of the set may cause community stress (misinterpretations, and hence discord).

Of course, greeting is only one of a multitude of sets of implications in the category "practices." It is true that all humans greet one another. It is also true that all humans eat, wear clothes, sleep, bathe, get married, raise children, worship God(s), educate their young, build homes, and conduct death rituals (funerals). In each instance your culture tells you how these should be done. Take eating, for example. American culture differs from Japanese culture regarding how this should be done. Americans eat beef and potatoes while Japanese eat fish and rice. Americans use knife, fork, and spoon while Japanese use chopsticks. Americans sit on chairs at a table; Japanese sit on the floor. Americans wear shoes while eating but not Japanese. Americans cook their meat; Japanese eat their fish raw. There are other differences as well. So, the question may be asked: Which culture is superior in knowing how eating should be done? The best answer, of course, is that neither is superior. They are just different. Arguing for the superiority of one's culture opens one to a charge of ethnocentrism. If someone accuses you of being ethnocentric, it is not a compliment. The "ethno-" element means culture or ethnic group. The "-centric" element indicates that you believe your culture is the center of the universe.

Many Americans are ethnocentric. We have a tendency to believe that (when it comes to questions of practices) the entire world should do things the way we do them. We believe everyone should speak the English language, use our measurement system even though most others prefer the metric system, and adopt our market and political systems. As one who has traveled extensively (to Mexico, Canada, England, Scotland, France, Belgium, Iceland, Luxembourg, Bermuda, the Soviet Union, Denmark, Sweden, Germany, and the Bahamas), I can report that other nations have indeed emphasized learning the English language and copying the American market and political systems. We seem to be gradually moving their direction on the metric system.

Before we berate Americans, claiming that Americans are ethnocentric, we should admit that ALL individuals have a tendency to be ethnocentric. There are ethnocentric Japanese, Germans, French, Arabs, Africans, etc. Considering our definition of culture, individuals in every culture have sets of implications. Assumptions do not require deep analysis. We often do not even give thought to what we are doing as we operate on a cultural level. Hence, we act ethnocentrically. The key to reducing community stress in this category is to actually give such matters some thought. Consider what the set of implications for the person from another culture may be. If you do not know, study. There are books at libraries and bookstores that will provide answers. You can go online and run a web search for the customs of the culture you wish to examine. Demonstrate to the individual from another culture that you are not ethnocentric, that you can appreciate the customs of the other culture. This should help to greatly reduce community stress.

In Chapter 1, I pointed out a few relief valves for community stress. Tyranny sometimes works and democracy sometimes works, but anarchy also works at times. In simplest terms, anarchy means, "nobody rules." I mentioned the nineteenth century motto promoting church unity that suggested: "In essentials, unity. In opinions, liberty. In all things, love." I pointed out that the second element of that motto is a principle of anarchy. This anarchy, which may be phrased as "In opinions, liberty," is precisely the relief valve I recommend for dealing with community stress related to our first category of cultural assumptions. Even the United States of America uses such a strategy to reduce community stress in a country that is a cultural melting pot. The Latin phrase on our currency, *e pluribus Unum*, contains a principle of anarchy (*pluribus*) combined with a principle of tyranny (*Unum*). America provides both unity (*Unum*) and plurality (*pluribus*). Burke writes:

> Let us consider for instance, the "principle" . . . that is implicit in the very name of our nation, which signifies a *plurality* acting as a *unity* (the pattern that is also quite accurately produced in the device, "*e pluribus Unum*"). As a union of states, we can accent our nation either as "The United *States*" or "The *United* States." The first accent would give us the Jeffersonian stress upon states' rights; the second would give us the Hamiltonian stress upon national federation. "Ideally,"

> as in the name of our country and in the pattern of its thoroughly accurate device,
> we can have both wishes (or "principles" in the first sense) at once.[30]

In order to have plurality, there must be some sense of anarchy. There must be some situations in which nobody rules. The suggestion that one avoid ethnocentrism implies that there is certainly room in cross-cultural communication for a plurality/anarchy principle. Why should people have to eat with knives and forks instead of chopsticks? Why should people need to sit on the floor to eat instead of on chairs at a table? Why should people be forced to eat potatoes instead of rice? These are certainly issues that allow for differences of opinions, plurality, anarchy. Are they not? Hence, when attempting to relieve stress that results from cross-cultural assumptions of our first category, it is advisable to study to understand how other cultures say things should be practiced. Then, you should demonstrate to individuals from other cultures that you have no problem whatsoever with their customs. You are trying very hard to learn other customs. This will produce much less likelihood of community stress.

Beliefs and Values. Our second and third classifications for cultural sets of implications are "beliefs and values." While the anarchy principle (that one avoid ethnocentrism) works quite well in our first category, it does not always work so well here. The term values indicates that cultures do place higher importance on some issues than others. No war was ever fought over whether to use chopsticks or spoons. Wars are fought when values are in conflict. Despite our strong political differences and our differing favorites among university athletic teams, my friend and I have never come to blows over politics or sports. There are values to which my friend and I and our families have given higher priority. Political pollsters tend to classify voters simplistically, based on their answers to series of questions relating to issues. But, human beings are much more complex than that.

As a doctoral student at the University of Illinois, I had the opportunity in a course on Rhetorical Philosophy to explore my own philosophy of decision-making. I hypothesized that humans make their decisions based upon a process that I called "systematically altered neutrality." It fits somewhere between the "methodological doubt" of Descartes and Wayne Booth's theory of "systematic assent." Descartes suggests that we doubt anything that may be doubted; whatever is left is truth. Booth proposes, "I will believe [in common sense] unless I am

---

[30] Kenneth Burke, *A Grammar of Motives* (Berkeley: University of California Press, 1969), 375.

given a *reason* to doubt."[31]   I personally submit that babies enter the world "neutral" on major philosophical questions. They know nothing of the "curvature of space" or even the spherical nature of the planet on which they live. To kids the earth, no doubt, appears to be flat. Nevertheless, they learn to accept, based on the word of their parents, friends, and teachers (some culture), the belief that the earth is round. Further study in logic, mathematics, and philosophy then casts new shadings on this belief.   Minds that were essentially neutral at first are systematically altered by each new argument, proof, experience, concept, or set of assumptions. We vacillate between doubt and belief, often finding ourselves back where we started--neutral.   But, by then it has become a different, informed neutrality.   In religious spheres it is called agnosticism (the position of not knowing). The agnostic is neither a believer in God nor an atheist. S/he claims not to know. S/he is neutral--neither doubting nor believing.

Some beliefs that might show up on a poll are held very shallowly, and therefore, while the statistic may be correct, there is not sufficient motivation for accompanying behavior. I may prefer fork and spoon to chopsticks, but I would not fight about it. What we call "values" may be those beliefs which we hold so strongly that, for all practical purposes, we could say that they are "set in stone." Such is the distinction between the political views that may have separated my friend and me, and the values that bound us together. True, we disagreed on whether or not the Indiana Hoosiers or the Fighting Illini should win the football or basketball game. But we have little discord concerning how life should be lived. The specific arguments, proofs, experiences, and concepts that brought us to our most strongly held common values were different, but we reached the same conclusion. We hold the same set of family values. If you polled America, you might even find that a majority of Americans agree with our views; but with us, these are not just "beliefs," they are "values." The difference that I described above applies. They provide motivation for our behavior! This is not to say that any of us are perfect examples of the behavioral set we view as correct. But one of the greatest motivations for our continued friendship is to be found in our sharing of these simple values. There is no question, for example, whether either of us would be found cheating on our spouses. There is no question whether we value our children and our families. There is no question whether we would ever consider divorce to be the best solution to marital difficulties.

---

[31] Wayne C. Booth, *Modern Dogma and the Rhetoric of Assent* (Chicago: The University of Chicago Press, 1974), 101.

We are certainly not naïve enough to think that all of our friends and acquaintances behave in the same manner. Perhaps that is one of the reasons that our friendship has been so persistent. It is difficult to find friends who hold these beliefs as "values." We have other friends whose company we enjoy, but with whom we have to tiptoe around issues that really matter to us. It's not that we mind tiptoeing. We do it regularly on political matters, for example, with each other, our families and friends. But these issues are our hearts and souls. I guess that everybody just needs a "soul mate" every now and then. Community stress is ever-present when humans interact, but community stress is minimized so long as the humans who interact at least share core values.

Anarchy worked in our first category. Why not advocate the anarchy principle as a way of dealing with conflicting cultural values? Let me offer some examples of values conflicts. Since I compared American culture with Japanese culture in the first category, I will compare American culture with Communist Chinese culture in the second category. Consider the different sets of assumptions these two cultures exhibit with regard to the value of baby girls. According to John Pomfret of *The Washington Post*, "Sex-selective abortions, infanticide and significant differences in children's access to medical care are contributing to an increasingly skewed sex ratio in China's countryside."[32] My recent web search for "infanticide in China" turned up two thousand four hundred ten (2,410) entries. Most of the infanticide and death due to reduced access to medical care is targeted at baby girls.

Certainly, infanticide is no new phenomenon. There is archaeological evidence of the widespread sacrifice of Phoenician children to deities dating to about 1600 B.C. Child sacrifice also occurred among the Mayans in Central America. In A.D. 315, the first Christian Roman Emperor Constantine tried to put an end to the practice of poor parents exposing their children to death. In 1858, a newspaper in Dublin, Ireland labeled infanticide in England as the greatest social evil. In 1922, England tried to stop the infanticide by enacting the Infanticide Act. Nor is American culture free from acts of infanticide. In 1994, Susan Smith admitted she drowned her two sons in a lake. In 2001, Andrea Yates confessed to drowning her five children in the bathtub.[33]

Yet, for American culture, this is not the norm. American culture is opposed to infanticide. This is not necessarily true in Communist Chinese culture. Since

---

[32] John Pomfret, "In China's Countryside, 'It's a Boy!' Too Often," *The Washington Post* (May 29, 2001), A01.
[33] "Infanticide: A History," *Newsweek* (July 2, 2001), 22-23.

1979, when China instituted a one-child policy to control population growth, widespread infanticide against females has occurred. *The Washington Post* reports:

> In heavily rural areas such as Guangxi, where boys are prized because of their value for farm work and because they inherit the land, the numbers approach 140 boys for every 100 girls, well off the international norm of 105 to 100 . . . . [R]esearchers are concerned that the trend . . . will reinforce a sense that the lives of little girls--and the women they grow up to be--are less valuable than those of boys and men.[34]

Bachelor villages are popping up. Girls from neighboring countries are kidnapped and sold to Chinese men. Girls are being turned into commodities. In Dalu village, on a plain near Guangxi's coast, seventy percent of the children are boys. Why? The real reason is economic: they need boys to fetch water, to guard property, to work in the fields, and especially to care for them in their old age. There is no government sponsored retirement program in China. Parents generally expect their sons to care for them. Additionally, China is patrilineal--inheritance passes to son from father. Women become part of their husband's family.

In the past, American culture may have also valued male offspring over female, but not to the point of widespread female infanticide. And even those earlier values have changed. Despite the fact that many argue that women do not receive equal pay in the workplace, the value of baby girls in American culture is virtually the same as the value of baby boys. We educate all children without discriminating. We anticipate the election of our first woman president. Employers are forbidden in our culture from discriminating based on gender. So, I will rephrase the question I asked in discussing our first category: Which culture is superior in knowing what the value of baby girls should be? Based on the earlier discussion of ethnocentrism, your first response may be, "neither is superior." But is that true? Do you really believe it is acceptable to kill baby girls so long as you live in China?

Try another example: The difference in value between the life of a Jew in Nazi German culture and in American culture during World War II. Nazi culture believed Jews should be exterminated; American culture taught that our Creator endows us all with the right to life. Which culture is superior in knowing what the value of Jewish lives should be?

---

[34] Pomfret, A01.

Try another example: The difference in the "right to liberty" values between the northern culture and the southern culture of America during the Civil War. Northern culture believed all humans were entitled to liberty. Southern culture disagreed. In retrospect, I think all American culture now agrees the northern culture, which prevailed, held the superior value. But, the earlier discussion of ethnocentrism would not have allowed such a conclusion.

Consider one more example. In many cities there is a disagreement between gang culture and law-abiding culture. Gang culture has no problem with drive-by shootings. The culture even endorses the practice. It is a value for gang culture. Not so for law-abiding culture. So, which is the superior culture on the issue of drive-by shootings?

Our second and third classifications of sets of implications--beliefs and values-- will not always accommodate an anarchy solution. The reduction of community stress often demands a tyranny principle. Some specific value set must rule. Otherwise, the community stress becomes intolerable. Authoritarian codes such as the Ten Commandments and the Constitution and laws of the United States and other countries are enforced. Yet, this is a principle of tyranny. Law enforcement officers and the courts of the land do not allow widely varying practices on these issues. There is a need for unity/*Unum*/United values on these "essentials." Government or culture becomes the tyrant. If America faces a community stress crisis in education, it is because it has taken an anarchy approach to values in the schools. The rash of school shootings in the late twentieth century gives evidence of such a crisis. Values are so often connected to religious cultures and, in an extreme attempt to separate church and state, the U.S. government has systematically eliminated values teaching from public educational settings. Clearly, more tyranny here would relieve community stress. This should not even require justification on religious grounds. Virtually all surviving cultures agree with many basic tenets:

- Don't kill!
- Honor your parents and teachers!
- Don't be sexually promiscuous!
- Don't steal!
- Don't lie!

Although these are admittedly found in the Christian, Jewish, and Moslem religions, they are also found in nearly every cultural code on Earth. Most people in America understand the need to be sensitive to cultural differences, but the baby has been thrown out with the bathwater, here. The reason my friend and I are able to handle *pluribus* in the areas of politics and sports is that we have *Unum* in other more important areas--our values. Even if you disagree with another individual on a specific political issue, you surely can find some bedrock values on which you agree. Build on those.

It is true. The values category of sets is where most heated arguments occur and where most literal wars are fought. This category of cultural stress, therefore, is the most difficult to relieve. Burke made it his motto: *Ad bellum purificandum.* Translated, his motto means: "toward the purification of war." Burke sought in language ways to purge the constantly recurring human impulse toward war (what I call extreme community distress). Chapter 2 discussed two linguistic ways of purging stress--tragedy and comedy. A third linguistic way to purge stress (which might function better in this situation) is dialectic. A favorite phrase of Burke's was, "Let the dialectic continue." Dialectic involves debating both sides of an issue. Some community stress will definitely occur as you debate deeply held values. However, the greatest community stress occurs when the debating stops and the fighting begins. The goal is not to *eliminate* stress; it is to *control* stress. Debate and discussion might purify or purge the war element from the community stress and, thus, keep the stress at "controlled" levels. Burke sees some hope for avoiding the holocaust, for example, in the fact that the political systems on earth cannot agree sufficiently politically "to set up the 'right' alignment and touch it off."[35] Hence, according to Burke, our inability to agree with each other may be our saving grace.

Of course, so long as it is unnecessary for two given individuals to settle issues of values among themselves, another relief valve is avoidance. There are times when political differences are best handled by avoidance. As stated before, sometimes I just changed the subject. I am sure you have heard that two subjects that are not talked about in polite society are politics and religion. Why is this? It is because religion and politics are the two greatest repositories of cultural values. Religion and politics tell us not only how things are to be practiced, but also especially what beliefs and values are prized. If the views given to us by religion and politics are "values," they are virtually "set in stone." People do not easily abandon their values. Sometimes, it is advisable to debate values in order to relieve community stress, as in the Chinese girls, Nazi, and Civil War issues. Sometimes, the debate is unsuccessful and full-scale war results. At other times, it is possible to live in separate cultures with competing values without needing to

---

[35] Burke, *Language*, 20.

settle the issues. This is one benefit of having cultures. You may retreat into the shelter of your own culture where you know people who share your values. This provides tremendous relief from community stress. It is not true that you should never talk about politics and religion. Sometimes, you may engage yourself thoroughly in political or religious discussion without any community stress whatsoever. This is when you find yourself surrounded by devoted members of your religion or political party.

All three classifications of cultural implications--beliefs, values, and practices-- contain issues of ethics. The subject of ethical stress will be considered more thoroughly in the chapter on conscientious stress. At this point, however, it is useful to discuss briefly the ways in which violations of manners contribute to community stress. Manners are closely related to ethics, but there is no strong moral component involved. The issue of manners pertains more to social status than to morality. In some cultures, slurping of liquids is the norm; in others, slurping is considered evidence of poor manners. In some cultures, belching is in bad taste; in others, a belch at the end of a meal is a welcome symbol of appreciation. It used to be considered mannerly in American business culture for a male to hold a door for a female. This is no longer necessarily true. It is amazing that others can interpret behavior that occurs with the noblest of intentions as thoroughly disgusting. It is no longer even safe to consult books on manners. Political correctness has made many of these books obsolete. By the same token, it is not safe to rely on political correctness; many people disagree strongly with political correctness. Slurping, belching, and holding doors are not major ethical issues. Yet, these issues can produce community stress.

Since all cultures are cooperatively formed, all living cultures are constantly in a state of flux. This is probably true now more than ever before in history. Transportation technology skyrocketed worldwide in the twentieth century. In America alone, people began the century riding in horse-drawn buggies and ended the century riding in spacecraft (with cars, trains, and airplanes in between). At the same time, communication technology has exploded--from the pony express to hand-held computers that can access a world-wide-web via satellite transmissions (with telegraph, telephone, motion pictures, radio, television, cell phones, and personal computers in between). Most of the dividing lines that were keeping cultures separate over the millennia are disappearing. Why is this important? The answer pertains to the way cultures are formed.

The definition of culture offered by O'Hair et. al. provides information on how cultures work, but it says nothing about how cultures are formed. Cross-cultural communication experts, Larry A. Samovar and Richard Porter briefly hint at the formation of culture in their definition:

> Culture is the deposit of knowledge, experience, beliefs, values, attitudes, meanings, hierarchies, religion, timing, roles, spatial relations, concepts of the universe, and material objects and possessions acquired by a group of people in the course of generations through individual and group striving.[36]

While this definition of culture supplies meat for the skeleton O'Hair et. al. had provided, the only reference to the formation of culture is that culture is formed "through individual and group striving."

I answer the question of culture formation by defining a culture as any communicating group that is in a sense "insulated" from other communicating groups. Humans, by nature, try to communicate with other humans. Most of our communication occurs among groups of individuals with whom they have been thrown together, insulated from other communicating groups. Several examples of cultural insulators are: language, geography, religion, philosophy, social class, gender, race, level of technology, age or milieu, family, tribe or nationality, occupation, level of education, physical constraints, politics, and family stage. Each of these insulators has a tendency to push individuals into one group or another. If you and fifty other individuals in your area speak English while the rest of the citizens speak French, you will likely form a culture with the other English speakers. You will get together and cooperatively determine your group's set of assumptions about how things should be done, what values are prized, and how people should behave. The same phenomenon would occur if you were geographically insulated from others--if you lived on an island with three hundred others, for instance.

But geographical insulators are fast disappearing. Transportation technology decreases the insulation of mountains, oceans, lakes, rivers, and deserts. Even language insulation is decreasing as countries become more educated. Being multilingual is becoming a norm, worldwide. Even the insulation of social class, once so inviolable in Europe and Asia, has become lessened in the American system. Americans may be upwardly mobile. They may climb from the lower class to the middle class to the upper class through wealth accumulation and education. Hence, there is more liquidity, less rigidity, in cultural norms. As the world becomes a global village, it may seem more likely that community stress

---

[36] Larry A. Samovar and Richard Porter, *Communication between Cultures* (Belmont, CA: Wadsworth, 1991), 51.

will be reduced because we are becoming more culturally integrated. This is not necessarily true.

The downside of this phenomenon is that you are faced with more cultural diversity than ever before. Rather than living within the low community stress environment of an insulated culture, you are thrust into a multicultural environment. You face many more disagreements over methods of doing things, values, and beliefs. While language and geographical insulators are decreasing in general, other insulators are not decreasing as much. Gender insulation, for example, is the number one biological source of insulation. Biology has not changed. Girls continue to be taught girl culture and boys continue to be taught boy culture. But there are new culture wars occurring in these gender cultures. The feminist movement has attacked many of the historic values of female culture. It has even wielded a strong influence on male cultural norms. The civil rights movement and forced integration have aimed at lessening racial insulation. In many ways, integration has successfully lessened insulation, but racial insulation continues to exist. And, forcing various racial cultures together has also increased the likelihood of more community stress. Religion and politics remain powerful insulators. Even some new insulators have developed such as new occupational insulators and level of technology insulators.

Small wonder, then, that you may be feeling more community stress, these days. You are simply faced with many more possibilities for culture-based disagreements than your ancestors who lived in insulated environments. Do not despair, however. Relief valves are available:

1. Study the cultures of those with whom you come in contact. Even somewhat outdated information is better than no information at all. At least it proves that your intentions are honorable.

2. Decide which values you hold that require cross-cultural unity. Must a respect for life, liberty, equal opportunity, etc., be present in society in order to avoid extreme community distress? Use dialectic (debate) to establish whatever values you determine are essential.

3. Seek out other cultures that might agree with you on those essential values. Begin to build relationships/coalitions with other cultures with values similar to your own.

4. Identify those ways of doing things and ways of behaving that are purely arbitrary, sheer matters of opinion. Allow for plurality and anarchy in these areas. Even demonstrate your willingness to try other ways of doing things.

5. Rely frequently on your own culture to provide an environment where the community stress is lessened.

6. Consider those areas in which your culture holds values that other cultures do not hold. Will society collapse if only your culture holds these values? If so, see relief valve number 2. If not, encourage others in your culture to keep them. This is much less stressful than demanding that all other cultures change. This is the principle of avoidance. When these issues produce community stress outside your culture, change the subject.

Responding to my view of the formation of cultures, a fellow graduate student at Purdue University, Michael Kent, offered me an interesting insight. Kent carried my insulation analogy to the next step. Since insulation is a building analogy, Kent urged that people "consider the R-Factor." The R-Factor refers to the cumulative total of insulation that exists in the walls, floor, or roof of a building. Kent was suggesting that communication would be increasingly more stressful the greater the number of cultural insulators that apply. Some insulators applied to my relationship with my friend. Certainly, we were geographically insulated and were not from the same family or tribe. Furthermore, we had different occupations and, as I have suggested in this chapter, we were politically insulated from each other. The vast majority of cultural insulators did not insulate us, however. We were of the same age and gender and shared the same language. We shared the same religion. Our philosophies were quite similar. We were in the same social class, the same race, and, with minor exceptions, we operated at the same level of technology. My kids' ages were nearly the mirror images of my friend's kids, so we were almost always at the same family stage. We had the same nationality, level of education, and physical constraints (corrective lenses for eyesight). Michael Kent is right. The R-Factor is determined by a combination of the number of insulators weighted in terms of how strong a "value" each insulator is considered to hold. My friend and I enjoy a very low R-Factor. Yet, the application of the relief valves mentioned above has been useful even in that relationship.

Chapter 4

Cash Stress

Cash stress is either (1) the stress you feel when you do not have enough cash to meet your wants and needs or (2) the stress you feel when you have plenty of cash and are not sure how you should invest it (although this second category is also a kind of confusion stress).[37] Those people who have plenty of cash but are not concerned about investment matters typically do not experience much cash stress. Most people with cash stress fit in the first category--not having enough cash. Even those who are serious investors are often investing for retirement--so that they *will* have enough cash to meet their needs and wants at a later time. Therefore, not much space will be devoted to the second category. If you have so much cash that you are only worried about how best to invest it, you can afford to hire an investment consultant. Good advice of that type varies according to the individual client and constantly changes--from month to month, day to day, hour to hour, and even from minute to minute. Hence, commenting in detail on that

---

[37] Burke, *Language*, 477, suggests as much when he states "persons of influence are beset by exceptional fear for their possessions."

type of cash stress in a book is a virtually worthless enterprise. Suffice it to say that the stress in such a situation pertains to two often-competing goals: investment return and safety. Those two concerns will be addressed below (in the section on retirement planning).

Cash stress comes for a variety of reasons. Sometimes, it comes through no fault of your own. In situations where the stress is not your fault it may be either no one's fault or the fault of someone else. As a financial consultant since 1979, I have witnessed plenty of examples of the type of stress for which someone else is at fault.

Someone Else is at Fault

Some occupations are fraught with uncertainty. It is hard to predict whether the specific position you hold will continue or not. A high-profile coach at a major university in Indiana was at the top of his field. He was elected into the basketball hall of fame. He seemed destined to become the NCAA coach with the most wins in NCAA history. His teams had won three NCAA championships. He operated one of the cleanest programs in collegiate basketball--no NCAA violations. His players graduated at a high rate. Academics were always a high priority in his system. He personally led campaigns that raised millions of dollars for the university library. Then, one day, a disgruntled former player charged that the coach had choked him in practice. Video of the incident in question was released to the media by a former assistant coach who had other disagreements with the coach. The video certainly did not clearly indicate an intentional choking attack, but the coach had placed his right hand somewhere close to the collarbone of the player. The media had a field day and a few months later the coach was dismissed. There is little job security in coaching. Although the specific coach in question was fairly well off financially, his assistant coaches were not equally well to do. They scrambled to find new jobs at the last minute. Even though the opponents of the head coach in question will argue that he was at fault for his dismissal, no one maintains that the assistant coaches were at fault for their cash stress. The fault belongs to the head coach, the media, the disgruntled player, the university trustees, or the university president (or a combination of the above).

Like, coaches, ministers are often dismissed without much justification. A youth minister I know had developed a sizeable youth program for a congregation. The teenage youth program developed from a total of six to about fifty teenagers in a relatively short time. The remainder of the youth program developed from zero to another fifty, for a total of about 100 youth during this period. However, some people in the congregation were offended that the program had grown without using traditional methods such as calling programs. The youth minister's wife had just become pregnant and the family had purchased a new home, but in a matter of weeks, they were without an income.

One possible relief valve for the coaching and ministry types of cash stress potential is avoidance. An old proverb says, "If you can't stand the heat, stay out of the kitchen." These professions are not desirable for those who would completely avoid cash stress. The security of the positions is frequently very low. The same is true for most new business enterprises. There is a high likelihood of failure in the first few years. By the same token, entrepreneurs and coaches (and maybe even a few ministers) who are very successful may well eliminate the first type of cash stress from their lives. Enduring some cash stress for a while may result in much lower cash stress later on. Walt Disney considered himself near bankruptcy three times before becoming a financial legend. Even cash stress can be a type of eustress. This is why some people gamble (both legally and illegally) on athletic contests. They do not enjoy the game as much unless there is some additional cash stress.

Every profession has its own versions of these cash stress stories with someone else at fault--personality conflicts with supervisors, economic downturns, a single customer's complaint, a corporate decision to eliminate a market, a desire to replace you with the boss's child, etc. In most instances, losing a job means immediate cash stress, and it may not have even been your fault. Other examples of innocent cash stress victims are those who have been victims of Internet credit card fraud, robbery victims, and those who have been taken by unethical contractors.

No One is at Fault

There are also periods of cash stress that occur when no one is at fault. Some people call events that cause such periods "acts of God"--as if God were at fault. God is blamed because it is assumed, *if He wanted to*, He could have changed history and natural events so that the cash stress would not have occurred. In Houston, Texas, a tropical storm dropped enormous amounts of rainfall in a brief time period. The city flooded in several areas. Houses and businesses suffered great financial damage. There were even many injuries and deaths as a result of the flooding. Houston is far enough inland that most would not have worried about flooding from a tidal wave and hot and dry enough that most would not have worried about flooding from a deluge. (Desert dwellers do not anticipate major flood damage either.) So, they may not have purchased flood insurance. Hence, cash stress results. My parents told me stories of the flood that hit their home before I was born. The floodwaters were supposed to have been retained by a levee, but the levee broke. Cash stress ensued. Even though some blame God for such events, the truth is that no one is really at fault.

If a huge earthquake hit Indiana, many Hoosiers would incur cash stress. Unlike Californians, Hoosiers do not anticipate major earthquakes; therefore, they do not own earthquake insurance. If a volcano erupted in Alaska, unprepared Alaskans would experience cash stress. I know that a few years back, many Floridian orange growers were financially ruined when a hard freeze not only killed their ripening crops but also killed the trees in their groves. A friend who lives near the coast of the Gulf of Mexico recently commented to me that some people have stopped building nice homes in a certain area along the gulf coast. Every few years, a hurricane comes along and the investment is lost.

This refusal to build in harm's way is an attempt to formulate a relief valve for such cash stress. Likewise, I have noticed over the past thirty years that Florida orange growers have moved further south. The citrus tower, which once provided a lookout point over orange groves in central Florida just north of the Walt Disney World area, now increasingly provides a view of residential developments, instead. In the Midwest, there are tracts of land that have been designated flood zones on which people are not allowed to build. Flooding has occurred on these tracts within the one hundred years prior and prudence advises against building there. There are indeed some relief valves for cash stress resulting from so-called acts of God. But there are also many possibilities for cash stress that would appear to be totally unforeseen.

You are at Fault because it was Preventable

Sometimes, those who have cash stress only appear to be victims. The cash stress, in a sense, was not really their fault. But, in another sense, it was their fault. They could have taken precautions. No mortgage company will lend money on a home unless the one who is mortgaging the home purchases homeowner's insurance. Insurance is the relief valve for many types of cash stress. Even homeowners who own their homes free and clear are wise to purchase homeowner's insurance. Houses might burn down or be damaged by high winds. Without insurance, tremendous cash stress might ensue. With insurance, the cash stress is greatly relieved.

Most state laws require car owners to own automobile insurance (at least liability coverage). If the car is financed, the bank or lending institution requires the purchase of full coverage. Driving a car without insurance opens the driver up to not only the cash stress of losing the value of the car but also the cash stress of potential lawsuits pertaining to any accidents. If the driver is fully covered and an accident occurs, some corporal stress may result, but the cash stress is relieved by the insurance benefit.

Many employers provide health insurance coverage for their employees and the employees' dependents. In those instances in which health coverage is not provided, it is important that the employee purchase a separate health insurance policy. One couple I know felt they could not afford to buy health insurance. Their son was involved in an accident that very nearly took his life. The physicians and hospital spared no expense to save the young man and, miraculously, he lived. But, the medical expenses were huge. The family eventually declared bankruptcy. Admittedly, bankruptcy itself is a relief valve for cash stress, but not a very desirable one. The cash stress would have been greatly diminished if the family had owned a medical insurance policy. If the deductible were as high as $1000.00 or even $5000.00, the cash stress would have been tremendously relieved. The family would have been wise to incur slightly more cash stress (the insurance premiums) for a while than to have undergone the devastating cash stress of bankruptcy. For a $2500 deductible medical insurance policy, the family of five could have paid as little as $130.00 per month in premiums. There are even short-term medical plans for those who have been laid off or are between jobs, those graduating from college who have not yet accepted an employment offer, dependents who are falling off their parents' health plan, and those who are waiting for group coverage. Even those who are considered uninsurable can generally get government-backed insurance. There is almost no legitimate excuse for not owning health insurance.

What about the worker who suddenly becomes ill or has an accident and can no longer work? It's called disability insurance. Most workers can purchase coverage that would replace about two-thirds of their incomes in the event of their disability. This coverage does not *remove* all cash stress due to the disability, but it does *relieve* the stress. Therefore, it is a *relief* valve for cash stress.

Crop insurance protects farmers from cash stress. Marine insurance protects sailors. Theft insurance protects your valuables. A young woman who rented an apartment near Louisville, Kentucky, saved a few dollars per month by not purchasing renter's insurance. Her apartment was burglarized and she lost what few valuables she possessed. She now knows that relatively inexpensive renter's insurance can relieve cash stress. Physicians insure against malpractice. Financial consultants insure against errors and omissions. Home contractors insure against liability for any damage resulting from their work. Real estate buyers purchase title insurance to protect their property claim in the event that someone else claims title to the land they have purchased. There are very few possibilities of financial loss that cannot be covered by some form of insurance. If you do not see the specific type of insurance you need advertised in your area, contact a major insurer such as Lloyds of London.

Some have commented to me that they are "insurance poor." They are spending all of their income on one or another type of insurance. They think it is wiser to not purchase insurance and take their chances. Perhaps this is valid. Not everyone's house burns down or is burglarized. Not everyone's car is involved in an accident. Not every doctor is sued for malpractice. Not every financial consultant is sued for errors and omissions. Not everyone's son is involved in a near-fatal accident. Not every renter is burglarized. On the other hand, some are. To live without insurance is to risk cash stress. Taking such risks is not dissimilar to gambling. Gamblers are constantly at risk of encountering cash stress. Owning insurance is a relief valve for cash stress. So many Americans faced so much cash stress due to the Great Depression that the United States government actually got into the insurance business--with Social Security.

Kenneth Burke calls activity similar to that engaged in by insurance companies the "socialization of losses." The socialization of losses could be extreme--as when everyone suffers together.[38] In a sense, this makes us feel better. There is a sense in which we do not mind going through the Great Depression so much because all of our friends are suffering too. Residents of San Francisco could endure the cash stress of a major earthquake because everyone there was suffering. They called it "earthquake love."[39] The socialization of losses can also be bureaucratized--as when nations turn to socialism. Burke describes how this is done: "One group . . . draws upon the *collective* credit of the government for support of its *private* fortunes, as when our federal treasury comes to the rescue of

---

[38] Kenneth Burke, *Attitudes Toward History* 3rd ed. (Berkeley: University of California Press, 1984), 67.

[39] Burke, *Language*, 460-461, describes San Francisco's "earthquake love" phenomenon as the "negating of the calamity and the negating of the rigidities that had arisen in the established [social] order." Thus, earthquake love suggests not only a relief from cash stress, but also a relief from the corporal stress of the calamity and a relief from the community stresses brought on by social classes. Everyone shared a common enemy--the earthquake. It became the scapegoat for all social ills and thus united all San Franciscans (Cf. also Burke, *Language*, 51).

the private banks. This policy for 'socializing' losses has been creeping into favor for many decades."[40]   Burke continues later: "Slowly this handy 'salvation device' becomes 'democratized,' as one group after another arises to claim its benefits."[41]   Most conservative Americans who oppose governmental socialism do not actually oppose the concept of the socialization of losses. They simply oppose the fact that the government *requires participation.*   Religious and community cultures were socializing losses long before federal governments began demanding participation in the socialization of losses. If someone in the church or community lost his/her home to fire, the neighbors and friends would take time out and help him/her rebuild. Such practices continue to exist, but to a lesser extent. Occasionally, we still see an entire community offer financial gifts to the uninsured victim of a terrible disease or accident. But, insurance has taken the place of much church and community socialization of losses. Like church and community socialization of losses (and unlike government-sponsored socialism), participation in insurance programs is typically entirely voluntary. Therefore, cash stress brought on by the refusal to purchase reasonable insurance coverage is the fault of the one experiencing the cash stress. The cash stress was preventable. The obvious relief valve is the purchase of reasonable insurance coverage.

A type of insurance that can be a relief valve for enormous cash stress is often overlooked--life insurance. One reason people are prone to overlook this relief valve is that the thought of death is too frightening. Just as trauma victims tend to "block out" memories of their traumas because they are too difficult to face, people tend to avoid all discussion of death. Over the years, I have counseled with numerous individuals who were facing death. Many of them were reluctant to discuss the matter even though death was immanent for them personally. Neither did their families and friends wish to discuss the death of their loved one. Some people refuse to go to funerals. The very idea of encountering death strikes terror in many. In *Persuasion, Proposals, and Public Speaking* (Say Press, 2004), I list the following facts:

1. Everybody dies sooner or later. . . .
2. There are expenses when people die (funeral services, casket, grave site, final medical expenses, monument, etc.)
3. Family and friends do not typically try to get by cheaply on these expenses when death occurs.

---

[40] Burke, *Attitudes*, 98.
[41] Burke, *Attitudes*, 161.

4. Not every person who dies has available cash to pay for these expenses.

5. Many who die have others who depend on them.[42]

Later in that book, I offer the following anecdote:

> Ten years ago, I proposed to my dad that he needed much more life insurance than he owned. Dad was in good health. He had no reason to suspect that he would die until he had reached a ripe old age. But what if he did? Mom would have to sell land quickly to pay the bills. Some of his children owed him substantial amounts of money. Could Mom prevail upon the kids to pay up at such a time? Yet, having loaned so much cash to his children, Dad did not have the cash flow at retirement to pay large insurance premiums. My proposal: Let's buy the insurance and have the kids who owe the money pay most of the premiums.
>
> Shortly after Dad qualified for the insurance, he became ill. The disease affected his kidneys. Soon, Dad was relying on dialysis. The procedures kept him alive for several years, but the inevitable day arrived. When Dad died, Mom and I placed a phone call to the insurance company. Without a moment's hesitation, they consoled us and said that all that was required was a copy of the death certificate. Within days, large sums of money came to our family income tax free. The proposal that I had made to Dad a few years earlier had solved the problem. Mom was taken care of. The debts the children had incurred to Dad were paid.[43]

Mom found out a few years later how difficult it is to sell your real estate in a short period of time. She took nearly a full year to sell her home and a few acres. Even then she received somewhat less than the real estate had appraised for. Had she been forced into a quick liquidation at the time of Dad's death, the financial loss could have been devastating. The purchase of life insurance is a tremendous cash stress relief valve.

You are at Fault

In some instances, the cash stress is brought on almost entirely by the unwise behavior of the one experiencing stress. A college president secretly siphons money from various funds in the college. He is fired. He and his family face cash stress. A vacation time-sharing organization deliberately sells more time slots than are actually available. The president goes to jail. His family faces so much stress, a divorce results. A wage earner develops a gambling addiction. The more

---

[42] Stan A. Lindsay, *Persuasion, Proposals, and Public Speaking* (W. Lafayette, IN: Say Press, 2004), 5.

[43] Lindsay, *Persuasion*, 107-108.

deeply in debt he goes, the more he gambles in hopes of climbing out of debt. The cash stress grows exponentially. A young married couple buys a new car, a home, furniture, appliances, decorations, and clothing--all on credit. Before they realize it, their payments are higher than their income will support. These individuals experience cash stress--but the fault is their own.

The best relief valve for the college president and the president of the time-share organization is abiding by the eighth commandment: "Thou shalt not steal." But, this type of relief valve will be discussed in more detail in the chapter on conscientious stress. It is amazing how frequently violations of moral and ethical codes result in cash stress in addition to the conscientious stress. It might even be said that the gambler is experiencing cash stress due to a failure to follow moral codes, as well.

I knew of a saleswoman who had a habit of meeting her prospects at a bar. She would get the prospect drunk and on the following day, the prospect would discover that he had purchased her product. There is enough blame in that situation to be shared by both the saleswoman and the prospect. The prospect claimed that someone else was at fault because the saleswoman had made the sale when he was not fully in control of his actions. The saleswoman claimed that all decisions were the choices of the prospect: to meet in a bar when sales would be discussed, to drink alcohol, and to purchase the product. Although, I agree that the saleswoman was unethical, I conclude that the prospect was at fault. We do not allow an individual to pass on the blame if s/he drinks alcohol, drives, and causes an accident. The responsibility for all actions resulting from the decision to drink belongs to the drinker.

Other than abiding by a good code of ethics, this variety of cash stress requires the relief valve called money management. Essentially, there are only two options for relieving cash stress through money management: (1) increase your income and (2) decrease your expenses.[44] The question to be answered is which of the two options should be chosen. There are a number of ways to increase income.

Increasing the income. The first way to increase income is to borrow additional funds. I have heard Christian financial consultants who quote Romans 13:8 out of context: "Owe no man anything (KJV)." They claim that God wants people to always be completely out of debt. If these Christian consultants would read further in that verse, they would observe that this verse is not talking about owing

---

[44] Aristotle makes this observation in his book *On Rhetoric* I.4.8: "for people become richer not only by adding to what they have but by cutting down expenses."

money at all. The only thing we should "owe" others, the verse says, is "to love one another." Although there are advantages to not having any monetary debt, sometimes, borrowing is the wisest course.

Before they attended college, my children held jobs that paid hourly wages close to the minimum wage. After graduating from college, their wages skyrocketed. A college degree makes a significant difference in one's earning power. So, we recommended to our children that they concentrate on their education first. We saw to it that their tuition and housing were provided. They were responsible largely for their meals, clothing, and incidentals. This meant that they had to borrow money to go to college. Since interest rates for student loans are typically very low, they began paying back their student loans when they were making the much higher wages of college graduates. It was wiser to pay back loans than to work at minimum wage to pay for schooling costs while attending college. Time is money. It took fewer hours of work to pay for the costs as a college graduate than it would have taken as a college student.

Generally speaking, if you plan to live in the same area for a few years, it is wiser to buy a home than to pay rent. Yet, rent does not involve borrowing money. Most people who rent have not saved up thousands of dollars so that they might pay cash for a home. Therefore, if they want to purchase a home, they must increase their incomes dramatically in a very short period of time. A mortgage loan is frequently a way of drastically increasing the income without drastically increasing cash stress.

Businesspersons, likewise, find that the wisest course at times is borrowing. The cash flow improves and the cash stress is reduced. Of course, except for certain types of student loans, all loans must eventually be repaid. The wisdom comes in knowing when the borrowing makes financial sense and when it is being used as a crutch to temporarily avoid cash stress resulting from poor money management. Salespersons must be able to demonstrate the long-term benefit of borrowing money right now. Loans should not be offered or used as a quick fix to support unwise impulse buying.

A second way of increasing income is to work overtime or find an additional job. This relief valve should actually reduce some cash stress, but it often increases other types of stress. A college student works overtime to reduce the cash stress of going to school. However, she now finds she does not get enough sleep. She is grumpy towards her roommate and sometimes misses class. She has less cash stress, but she has more corporal stress. She needs those hours of sleep she is missing. Her roommate does not passively accept her grumpiness. The two argue. She has increased community stress. She has fewer hours to work on her class projects. She is running up against deadlines. She experiences increased chrono stress. Working more hours is not always a successful way of reducing overall stress. However, when one discovers that he has a great deal of leisure time, but cannot afford to do anything with it, working longer hours may make a great deal of sense. Not only is the income increased, but the expenses (related to leisure activities) are also decreased.

Getting a raise at work or taking a new position with better pay increases income and generally does not increase other types of stress. One exception to this rule may be that you experience greater community stress when you need to ask your boss for a raise. Another exception to the rule may be that the new job or raise brings with it additional responsibilities. You may again experience increased chrono stress as described above. Furthermore, if you feel somewhat incompetent in handling your new responsibilities, you may experience more competence stress.

One of the most clearly stressed individuals I have known had just received a fantastic promotion. The pay was much higher than he had been receiving. His superiors were favorable to him in every way--hence, the promotion. He was scheduled to take some vacation time, so the chrono stress would appear to have been quite low. Yet, the competence stress brought on by this promotion had so pumped his stress balloon that he lived constantly on the verge of explosion--in great distress. A couple of years later, he changed his position to a slightly lower rank although his pay remained the same. The competence stress was greatly relieved. The relief of cash stress is not the singularly important goal of stress management. Navigating the seven C's of stress demands that you take into account all seven stressors and how changing the effect of one stressor affects the other six.

Decreasing the expenses. The most effective relief valve in money management for cash stress may be to just decrease the expenses. This does not require increasing chrono stress, competence stress, community stress, or corporal stress. The steps are easy to take:
1. List all expenses you currently face.
2. Calculate your options. Which expenses are unavoidable? Which expenses can you reduce by selling your current asset (car, home, boat, RV, etc.) and not replacing it? Which expenses can you reduce by selling your current asset and replacing it with a lower priced asset?

3. Prioritize your expenses. Start with all unavoidable expenses. Then list in order of importance your additional expenditures. Begin by eliminating (temporarily) those assets you could afford to do without and by listing the lowest reasonable price for those assets that could be replaced.

4. Calculate your total available income after deductions.

5. Now, budget. Fit the top priorities into the budget first. Then, see if everything else will fit after eliminating and replacing some assets. If there is any income left over, this is your disposable income. You may now return to those assets you eliminated and replaced and see which, if any, may be kept (based upon the priorities you had set).

Sometimes, unforeseen financial crises occur that can disrupt even the most reasonable budget. I will cite my own example. The difficulty occurred at the height of my business career. I was making plenty of money and facing very little cash stress. A couple of years earlier, I had won a $15,000 European cruise-for-two from an insurance company I represented and, before accepting such a trip, I had checked into the taxable nature of taking the cruise. The IRS handbooks had indicated that "educational travel" for educators was tax-deductible; so I spent the majority of my time during this cruise studying and researching. I did research for my book *Revelation: The Human Drama*, which was eventually published in 2001 by Lehigh University Press. Since I taught cross-cultural communication elements in my courses at the college level, I considered the visiting of various Scandinavian cultures educational travel. The IRS publications seemed to indicate that 100% of my expenses for the trip were tax-deductible. I chose not to claim the entire deduction since I had obviously also enjoyed seeing the sights of Denmark, Sweden, Gotland, Germany, the Netherlands, Finland, and the Soviet Union. Trying to be fair, I deducted 1/2 of the expenses.

To begin this later year, the IRS threw a tax-audit party for me. Although I had never been through such an ordeal before, I thought I could handle it myself. I didn't hire an attorney or a CPA. As intimidating as a tax audit can be, I found that I satisfied the IRS auditor with regard to just about everything in my report-- except the cruise. I sat with our auditor and argued our case. The example given in the IRS publication was that a "French teacher" could take an extended vacation to France and do nothing but talk and absorb French culture and language during the visit and it would be 100% tax-deductible. I observed that I taught foreign languages as a college professor. I also pointed out that my research language in my Ph.D. program at the University of Illinois was German, which I was indeed strengthening in my journey through the Scandinavian countries, if she needed to have an exact parallel with the "French Teacher" example.

Ultimately (and foolishly), I expressed the fact that I was willing to trust her judgment and accept her ruling. I was too intimidated to follow the appeals process. She disallowed my deduction. With taxes, interest, and penalties, they eventually wanted me to pay several thousand dollars more to them in taxes. I did not know how to deal with the IRS. An accountant later told me I should have appealed. Even if I lost, he said, they would have settled for about $1000.00. But I had no time that spring to fight the IRS. I agreed to their terms, confident that I would be "money ahead" by concentrating on business and settling with the IRS. I was making good money!

I attended a "Leader's Forum" conference staged by the company that had awarded me the cruise in March of that year. I was ranked as the top college market producer and the top college market manager. I felt confident enough to lobby the president of the company to argue for a stronger commitment to the college market but I may have embarrassed him in front of the new German owners of the company. I thought I detected a hint in his words that the company was becoming less interested in the college market so I asked him if that was the case. His answer sounded ambivalent.

One month later, the company began dismantling their college market. A year later, they were entirely out of the college market. But, as a man who had been making a very nice income in that market, who had just made extensive improvements to his already expensive home, who had just thrown in the towel in the IRS battle to the tune of a few thousand dollars, I had made a nearly fatal mistake. I had been too loyal to a single company. I hadn't diversified. I had violated all of the clichés. I had hitched my wagon to a single star. I had placed all of my eggs in one basket. I swore to myself, "Never again!" The only other company with which I had been placing any serious amount of business became the base for my backup plan. I had been placing my annuity business with them. With that base, I went about the business of surrounding myself with twenty top-notch companies, refusing to ever again become the victim of a single company's corporate whims.

It took three years to rebuild my business to the level at which it had been. During that time, my wife and I engaged in heavy-duty money management--decreasing expenses. We sold every asset we could live without; however, we refused to *give* our assets away. We did not receive a reasonable offer on our house, so we kept it. But, at least, we were willing to list it. We were willing to live in a lower priced home so long as the exchange actually helped us relieve cash stress. I borrowed money to put my business back on its feet. I liquidated my retirement account. I worked longer hours. I pursued every relief valve available to me.

When I reemerged from my crisis, I was determined to avoid similar situations in the future. The IRS behaves like a shark when it senses blood in the water. I learned that the IRS does not operate honorably. When an agent suspects that a taxpayer will go along with the agent's unfair decision, the agent will take the taxpayer to the cleaners. So, I hired an attorney. I sought the advice of a CPA. I fought the IRS. I followed the appeals process . . . and won. I could finally see that my attempt to save money by not hiring professional consultants in the first place was penny-wise but pound-foolish. I will admit, though, that I learned a great deal about cash stress management from this ordeal. I learned to have back-up plans. Cash stress can result from unforeseen crises. Back-up plans can be important relief valves.

The final cash stress I want to discuss is the futuristic one. I refer to the cash stress retirees feel. Most wage earners have at least forty years to accumulate funds for retirement. Yet, the vast majority of wage earners do not accumulate enough. According to Social Security statistics, out of one hundred wage earners who begin working at age twenty-five, thirty-three will die before reaching age sixty-five. If you are one of those thirty-three, you will not have cash stress at age sixty-five because you will be dead. I think most of you will agree with me, however, that retirement expense is one cash stress you would prefer to have (as opposed to the alternative).

If one-third of the beginning workers die before retirement, two-thirds remain living and therefore need retirement funds. That means sixty-seven out of one hundred will live to retire. Let's hope you will be among those sixty-seven! If you are, statistically five of the sixty-seven will still be working at age sixty-five. They may feel they cannot afford to retire but at least they are able to continue working at their job. Four of the sixty-seven will be able to retire well to do financially. One of the sixty-seven will be wealthy. You would be surprised how many people I have consulted with who plan to be that one out of a hundred who is living and wealthy at age sixty-five! If you cannot be wealthy, you would probably settle for well to do. But the one group you should avoid like the plague is that remaining group of fifty-seven out of sixty-seven who reach retirement age flat broke. This is why retirement groups become hysterical whenever one politician accuses another politician of causing harm to the Social Security program. Social Security is the only form of retirement planning most retirees have. They would be in impossible straits without it. These retirees are in a state of extreme cash stress whenever they conceive of losing Social Security benefits. But, of course, Social Security recipients have less cash stress than those who were broke at retirement age before Social Security was instituted. Social Security itself is a relief valve.

Imagine, however, how much less stress those who planned for retirement since they began working face. If a person who begins working at age twenty contributes only ten dollars per day throughout his lifetime to a tax-sheltered retirement program earning an average of seven percent per annum interest, s/he accumulates over one million dollars by age sixty-five. If the same person waits just five years to start planning for retirement (i.e., until age twenty-five), s/he needs to contribute fourteen dollars per day to accomplish the same feat. If s/he waits until age thirty, she needs to contribute twenty dollars per day--twice the daily amount of the person who starts at age twenty. If s/he waits until age forty, the amount needed jumps to forty-two dollars per day ($15,000/year). If s/he waits until age fifty, daily investment required to achieve millionaire status is one hundred five dollars per day ($38,000/year). Needless to say, a lot of fifty-year-olds wish they had begun planning for retirement at age twenty. The first key to relieving cash stress at retirement is to start investing immediately.

But in what should you invest? A good investment plan could include six steps. *Step One* is to lay the foundation for your future investments. Create an immediate estate so that, if you die, your family will be taken care of. This is accomplished by purchasing life insurance, but I recommend purchasing the type of life insurance that offers both minimum guarantees on the earnings of the cash values plus current competitive rates. Two products offer both of these elements: universal life insurance and interest sensitive whole life insurance. Term life insurance is not an investment vehicle; it is only a protection vehicle. Variable life insurance does not typically guarantee the investment principal. The investor is risking too much at this stage by purchasing variable life insurance. All of the investment could be lost in a severe economic downturn.

Step Two involves maintaining at least three times your monthly income in an insured savings institution like a bank, insurance company, or savings and loan. The interest rates may be lower than you could find at some other institutions but the principal is guaranteed and your money is more liquid--you can get it out of your account or annuity at a shorter notice, if you have emergency needs. I typically recommend putting most of your retirement funds in an insured annuity.

Safety of principal is more important than rate of return when it comes to retirement funds.

Step Three is the purchase of a home. Home ownership is the American Dream, but you need to be careful here. There was a time real estate values climbed rapidly throughout America. This is not necessarily true at all times. In some locations, real estate values may be on a downward trend. I heard recently of an individual who built a new home. The construction costs were $200,000. Three years later, he sold the home for $125,000. He had built a home that was too expensive for the neighborhood in which he built. There was a time all mortgage interest payments were tax deductible so it was at least financially wise to pay interest instead of rent. This is not necessarily true any more. Federal income tax calculations provide a "standard deduction" in lieu of itemizing tax deductions on Schedule A. Potential homebuyers should calculate their total Schedule A deductions to see if the home purchase would actually produce more tax deductions than the "standard deduction." If more deductions result, how significant are the additional deductions?

Whatever you do, avoid buying more home than you need in order to justify purchasing a home. Do not buy one that costs $50,000 more so that you will have more deductions. Keep in mind: renting is usually more of a liquid financial arrangement than buying. Does the possibility exist that you will be relocated? Will you change companies? Will the new company guarantee the fair market value sale of your home? You could jeopardize your financial portfolio by purchasing when you should be renting. But, if you feel confident that you will be living in the same area for a number of years, home buying may make a good deal of sense. Renters do not typically set aside the difference between the monthly rent and a house payment in an investment account. Mortgage payments, on the other hand, do serve as a type of forced savings. You build equity and indicate that you have some financial stability.

Be very careful, however, not to overpay for your home. If you resell it for the same amount you paid for it, you will probably lose money. The realtor's fee is typically five to seven percent of the sales price. If you buy a home today for $100,000 and sell it tomorrow, it must sell for at least $105,000 to $107,000 for you to break even, if you use a realtor. If you purchase the right-sized home in a neighborhood where property values are increasing, pay a fair market value for it, and remain in the home for several years, your home purchase stands a good chance of being a wise investment.

In *Step Four*, you may want to begin investing in mutual funds. This is the first stage in investing in the market. Mutual funds have the advantage of being diversified. Mutual funds diversify their risks by investing in several corporations. If one company encounters a downturn, other growing companies may balance the loss with their gains. Managers of mutual funds are professionally trained to handle your money. Even so, I advise against putting the majority of retirement funds in this type of vehicle. This investment should be considerably smaller than the investments you make in steps one through three.

I do not recommend moving to *Step Five* unless and until you have solid assets in steps one through four. This step involves blue chip stock investment. Blue chip stocks are old-line companies that have a history of paying consistent stock dividends. Nevertheless, even blue chip stocks can go under. Consider Enron, MCI, and Firestone. Do not invest as much here as you invest in the safer mutual funds.

Only when you have extra cash that you can afford to lose should you move to *Step Six*. This step involves investing in speculation. Here you may experience big losses or big gains. Since you could lose everything you invest in this area, consider how much cash stress you are willing to accept before investing. A small investment that turns into a huge gain is a very happy situation. A small investment that is lost completely may not produce undue cash stress. Just be careful.

Remember, cash stress is either (1) the stress you feel when you do not have enough cash to meet your wants and needs or (2) the stress you feel when you have plenty of cash and are not sure how you should invest it. If you have advanced to step six, you have reached that second type of cash stress. My suggestion: Play it smart. Keep your retirement funds safe and only take chances on that which you can afford to lose without suffering cash distress.

Chapter 5

Chrono Stress

One of my students exploded at me following a quiz he failed. "I'm working forty hours per week, plus taking a full load at college, plus trying to find time for my family!" he screamed at me. "Plus, I'm not as young as these other students; I can't remember all of those things you're asking on quizzes!" Here was a classic example of distress caused by working too many hours. Chrono stress is the stress one experiences when there seems to be insufficient time to accomplish one's tasks. I said to my student, "First of all, calm down." I explained to him that university level work typically requires two hours outside of class for every hour spent in class. Since he was taking sixteen hours of coursework, he should expect to spend thirty-two additional hours preparing for his courses each week. This means his full load at school will take forty-eight hours per week. Since he is working forty hours per week, he has committed to eighty-eight hours of work and school. If he sleeps eight hours per night and takes one and one-half hours per day to eat and use the bathroom, his total hours used each week amount to one hundred fifty-five. If he drives one hour each day to and from work and classes, his total is one hundred sixty hours per week and he still has not seen his family. There are only one hundred sixty-eight hours in a week, and he has committed himself for virtually all of them. His problem is not the difficulty of my quizzes--others do quite well on them--his problem is chrono stress. He is trying to pack too many things into each day. His stress balloon is expanding and he explodes at small irritations. He needs a crash course in time management.

There are four basic steps in time management:
1) List all tasks to be performed.
2) Prioritize these tasks.
3) Allocate time for performing each task.
4) Schedule the tasks, beginning with the highest priorities.

It is impossible to manage your time unless you first identify those tasks you wish to perform. What financial tasks are there? What personal tasks? What household and family tasks? What job-related tasks? What church-related tasks? What recreational tasks? What essential tasks? How much sleep do you need? How many meals? How much time for showers and bathroom breaks? How much time for relaxation and entertainment?

Once the tasks have been listed, it is necessary to prioritize. Certainly, everyone needs sleep, so this is a top priority. What other top priorities are there? Religion? Family? Food? Shelter? Position those tasks, which have been listed, in the order of their priority. When paperwork comes across my desk, I prioritize it. I have an "urgent" slot, a "when I get around to it" slot, and a "possibly" slot. I rarely "get around to" the second slot and almost never do a "possibly." But the "urgent" gets handled.

After you prioritize your tasks, you may try to increase the extent of accomplishments by utilizing timesaving strategies. Perhaps, less-qualified individuals whom you might be able to hire without producing an unacceptable amount of cash stress can perform some tasks. Perhaps, while driving, meals can be consumed, cellular phone calls can be placed, educational cassettes can be listened to, and news reports can be absorbed. Perhaps an investment in a microwave oven, a computer, a plane ticket, or a riding lawn mower makes sense in order to reduce chrono stress.

But ultimately, you simply must come to grips sooner or later with the fact that you cannot do everything you are interested in doing. You must be willing to do whatever you can and be satisfied with that.

Time Management and Life Planning
I had moved from the University of Illinois to Purdue University because I saw

the opportunity to teach for Milligan College in their extension at Purdue while I also built a business. However, I moved before I should have. My academic credentials were incomplete. I had completed the Ph.D. coursework and two of three preliminary exams at the University of Illinois before I moved. My advisor and others on my doctoral committee at Illinois then moved from the University of Illinois so I had to find new members for my doctoral committee. New members of my committee then started requesting additional work from me before I took my final preliminary exam and began my dissertation work. It was necessary for me to drive back and forth to Illinois. Furthermore, I had devoted much of my time to earning a living in Indiana. I did not have as much time as I had anticipated to devote to completing my Ph.D. at Illinois. I was experiencing chrono stress. My limited time period for finishing the Ph.D. at Illinois was rapidly running out. I asked for time extensions so that I might complete the Ph.D. at Illinois. I thought they owed this to me since I had completed all doctoral coursework at the University of Illinois with a perfect GPA, but they concluded that too many members of my committee had moved away. I would just have to complete the Ph.D. at another school.

I checked into other graduate schools and considered opting for a program that was sponsored by the University of California at Berkeley. I was accepted, flew to Berkeley to discuss my plan of study with professors there, and agreed to a certain plan. I was teaching part-time for Purdue's English Department at that time, plus running a successful business, plus coaching my kids' basketball and baseball teams. So why not dive headlong into a Ph.D. program on the other side of the country as well! This plan was not well conceived. There simply was not enough time in my schedule to do everything I wanted to do. I found it necessary to do some time management.

I listed my tasks: eat, sleep, run my business, teach at Purdue, spend time with my family, teach at my church, maintain my house and grounds, travel the world, take graduate coursework, and pay bills. Then I prioritized. First, I knew I needed to sleep. Second, I needed to eat, but that could be combined with my third priority--spending time with my family. That was a very high priority for me. For my fourth priority, I had to decide whether my business or my Ph.D. was more important. The Ph.D. won but I still had to pay the bills, my fifth priority. Traveling the world was fun but it was something I could put off until later. Teaching at church was a very high priority, but I had done this teaching completely gratis in the past. There were churches that would pay me to teach and, thus, help me pay the bills.

Then I began to allocate time for each task. Sleeping would consume fifty-six hours per week. Eating and spending time with my family demanded at least three hours per day or twenty-one hours per week. My Ph.D. coursework required fifty hours per week. Teaching at church required seven hours per week. In order to receive free tuition for my doctoral work as well as extra income to pay the bills, I taught courses at the university nine hours each week. This left only twenty-five hours each week for my business and everything else. Fortunately, I had built up renewal income in my business. I could expect to receive a certain level of business income without working at my business. I knew that income would eventually run out if I did little to expand my business, but I had already set my priorities. My business activities would be necessarily curtailed.

I planned my schedule. I budgeted time for sleeping, eating and family first. Then, I budgeted time for my Ph.D. coursework. I fit time to teach for the university into my schedule, then supplemented that teaching time with salaried teaching time at church. The time slots that remained were given to business activity. I found that I needed some additional financial help in the form of student loans to avoid facing severe cash stress. I also knew that after two years the bulk of my renewal income from my business would run very dry. Therefore, I was determined to finish my Ph.D. program in two years. No one in the history of Purdue University's Department of Communication had ever accomplished that task in so short a time. But, I did--by using time management principles. My major professor at Purdue, Don M. Burks, would later comment, "Together, then, though the effort was of course very largely Lindsay's, we brought about the most expeditious progress toward the Ph.D. degree in the long history of Purdue's Communication Department. What's more important, Lindsay's dissertation is rigorous in its scholarship." When the renewal income began to run out, I was finished with my Ph.D. I began a teaching career and found that I had to engage in new forms of time management. But that is another story.

Time Management and Vacation Planning

Chrono stress can be present even you are not working. It even occurs when you least expect it--when you are on vacation. Before I began my doctoral studies at Purdue, I made plans to travel out west for a couple of weeks. On this trip, my family invited some friends to travel with us. In order to contain the costs of making a long trip to California, my wife and I had purchased membership in one of those travel club deals. I finally broke down and bought a vacation package at a sales presentation. We were promised unlimited low-cost lodging at vacation destinations everywhere. They told us our friends could stay with us at no charge. I actually surprised myself by agreeing to the purchase after the sales presentation. I reasoned that I would use the resort accommodations during our California trip and save money in the long run. Our friends and we met and planned our itinerary. We assumed that the accommodations at the various resorts along the way would be of a quality similar to the resort where we had purchased membership.

That mistaken assumption was dispelled on the very first night of our trip. We had not realized that every one of the resorts would be at least an hour out of our way. This produced chrono stress. We kept wandering through the winding back roads, following the signs to our resort destination. Upon our arrival, we were directed to an old dingy travel trailer--our resort accommodations. We all expressed the hope that this was an aberration. Surely the other resorts will be better, we told ourselves. We took a dip in their moderately dirty and run-down swimming pool and drove another half hour to the nearest town to eat pizza.

We returned to our "resort," got some sleep, and started on our journey again the next morning. After we had been on the road for an hour, my wife noticed that she did not have her purse with her. No question about it, this was going to be a typical Lindsay vacation. The last time she remembered having her purse was when she went into the pizza parlor the previous night. We found a phone and called the pizza parlor. They had the purse. We retraced our path for another hour, and then, after retrieving the purse, we raced toward our next "resort" destination. We felt more chrono stress.

Due to the two hours lost in retrieving the purse it was late when we pulled into Tucumcari, New Mexico. We arrived at our "resort" motel and choked back the tears. It was an old run-down structure. But we were all tired. We had reservations for king-size beds rather than the cramped travel trailer bunks of the previous night. We inspected the rooms. Upon walking into the first room we noticed huge bugs on the bed and floors. We decided not to take the rooms but as we gathered to load up and drive away we noticed that some of the kids were missing. They had assumed that we were staying and had made themselves comfortable on the beds in one of the rooms. We quickly gathered them, straightened the beds, and drove to a Motel 6--a truly luxurious alternative, compared to our "resort" reservations.

Our plans for the next day were too ambitious. We wanted to see the Petrified Forest and the Painted Desert and still make it to our resort, sixty minutes south of Flagstaff, by evening. Chrono stress hit us once again. We had not budgeted enough time to thoroughly explore the sights. Upon our arrival, the beauty of the landscape became too much of a temptation. The "hiking spirit" possessed the crew. The natural wonders of the national park were best viewed by following trails through the cliffs, hills, peaks, and valleys. There is no question that the adventures were fun and worth the time spent but it was after midnight as we negotiated the dark winding mountainous roads which led to another secluded resort setting.

We had paid for the accommodations in advance and had been required to spend a minimum of two nights at the resort. As we pulled into the resort area, we looked for security guards. There were none. Anyone could come and go as they pleased. We found a booth where we assumed that the security guard would leave us a message regarding our pre-paid reservations. The booth was locked. There were no messages. We wandered throughout the resort in search of anyone who could help us. No one was awake. Our only companions in the darkness were bats, which would occasionally swoop down toward us. We located an unlocked card room but the windows were closed and the interior temperature of the room was high. I suggested that we camp out on the floor of this room but the others were not so inclined. We parked the two vehicles and people tried to catch some sleep in the vehicles. After an hour of discomfort I followed my own advice and moved to the card room. At last, I fell soundly asleep and managed to stay that way until the early strains of daybreak. The others had endured the discomfort of sleeping in the car and van. We all took showers and cleaned up in the camping resort's bathroom facilities and awaited the arrival of some security guard.

At 8 a.m., he arrived. I complained concerning our lack of accommodations. He apologized, refunded my pre-payment, and we left our third resort destination with a strike three count, out of the first three pitches. Our original plan had called for us to spend the next day at the Grand Canyon before returning to the same resort in the evening. Obviously, that plan would be revised. We were certainly not returning to that resort. At evening, we found motel accommodations in Kingman, Arizona. Through the desert and on to southern California we drove. We hoped that at last we would find a decent resort at the end of the trail.

I learned something on that trip--that vacations are intended to be relief valves for chrono stress, not stressors themselves. By performing detailed time management on a vacation, we had managed to make every leg of our journey a stress-producing experience. In an attempt to reduce cash stress by booking our lodging in advance at "resorts" along the way, we had increased chrono stress. Since that trip my wife and I have tried to be less precise in our agenda planning for vacations. We go where the road takes us and stop when the sun goes down. Of course, there are always situations in which lodging is difficult to find but, generally speaking, vacations should be used as chrono stress relief valves, not stressors.

Time Management and Business Planning

The business document that best indicates time management is the Planning Report. In my business writing courses, I require business students to follow virtually the same four steps mentioned earlier to produce a planning report. I give this assignment as the first phase of an overall business proposal assignment. The goal of the planning report is to:

1) List all tasks to be performed,
2) Prioritize these tasks,
3) Allocate time for performing each task, and
4) Schedule the tasks, beginning with the first priorities.

I recommend that my students organize their business proposals around six phases--problem identification, problem proving, causal analysis, solution analysis, proposal development, and persuasion. Since the term "proposal" is shorthand for the phrase "proposed solution," some type of problem is implied. Why else would a "solution" be needed? Therefore, in their planning, my business students begin by listing tasks they will need to perform to identify possible problems within a company. For example, they may need to do field research such as on-site observation, interviews (with owners, managers, employees, customers), and/or questionnaires. They may need to do library research. They may list tasks such as locating and studying journal articles pertaining to the type of company they are assisting. They may need to brainstorm--coming up with possible problems their business might encounter. Then, they need to analyze the various problems they have identified. What problems are most easily demonstrated or proven? What problems are having the most adverse effects on the business? What problems are getting worse? Are there any problems that are clearly shown to be a negative trend? It is not necessary in a planning report for the students to actually choose a problem to solve. All they are doing here is time management. They are simply determining how much work will be involved in this project. They are still at step one--they are listing tasks to be performed. And, they have only begun to list tasks.

Once the students have identified a specific problem they wish to solve, they proceed to the problem-proving phase. Here, the tasks to be performed include such things as securing documentation. What reports, studies, accounts, etc. will help us prove statistically that the problem we have identified is worsening? Where can we find statistics that will indicate when this worsening trend began? Is this problem much bigger than the specific company we are assisting? Can we do Internet research or library research to demonstrate that the trend is affecting an entire industry? All industries? If so, when did the mega trend begin? Are there authorities, experts, whom we can interview? Will they be able to corroborate the statistics we have uncovered? Are there any anecdotes or case studies that will lend a sense of real life to the statistics? With whom would we talk in order to learn the details of such stories? Again, in the planning report, the goal is not to secure answers to all of our questions. The goal is to list the tasks we will accomplish in finding such answers.

As students consider the next phase, they list tasks associated with causal analysis. They may want to brainstorm--to form lists of various types of causes that may have produced the problem they have identified. What causes may have occurred at precisely the time the negative trend began? Such causes are called *immediate* causes. What causes have always been present? Such causes are termed *remote* causes. What causes may have continued or increased the trend after it began? Such causes are *perpetuating* causes. What causes are absolutely essential for the trend to have existed? (The trend could not have occurred without them.) Such are *necessary* causes. What causes could have produced the trend all by themselves? These are *sufficient* causes. What causes are *obvious*? Which ones are *hidden*?[45] Besides brainstorming, where can we find such causes? Are there articles in journals pertaining to such matters? What do the business owners say about the causes? What do the managers and employees say? Do we know others in the same business who may have suggestions as to the cause? Once the students have identified and classified all possible causes, they must analyze the importance of each cause. They are looking primarily for a cause or two that are immediate, necessary, and sufficient. Perpetuating causes may also be important. However, they can rule out remote causes. Remote causes do not explain why the trend began when it did because they have always been present. Once again I remind my students that their job in a planning report is not to answer all questions; it is to determine the tasks that must be completed in order to find the answers.

---

[45] For a fuller discussion of these causes and how to use them to do causal analysis, see Stan A. Lindsay, *Persuasion, Proposals, and Public Speaking* (W. Lafayette, IN: Say Press, 2004), pp. 15-17, 99-103.

Upon listing tasks to be performed in the causal analysis phase, the students may turn to listing the tasks to be performed in the solution analysis phase. Here, they must list all possible solutions to the problem--both good and poor. They may want to do field research and/or library research again. They certainly will want to brainstorm. After they have listed possible solutions, they must analyze those solutions. Which solutions address the cause of the problem as determined in the previous phase? Which do not? Which solutions will work? Have the solutions been tried anywhere? Where? Must we do further research with regard to case studies? Did they work? Which solutions are feasible and which are not? Can the business afford to implement the various solutions? How easy will it be to implement each solution? What tasks must be performed in order to answer these questions? Do not answer all questions at this point; just list the tasks.

Once solution analysis is completed, the students can begin to develop their specific proposal. They will choose only elements from the previous phase that meet three criteria: they address the cause of the problem, they will work, and they are feasible. Additionally at this phase the students need to consider how detailed their specific proposal must be. Will the leaders of the company understand exactly what they should do? Will someone in the company be able to implement the proposal as they provide it? Or, must the students include in their proposal outside assistance? Such questions must be answered when the time comes, but not now. The students are still in step one of the planning report-- listing tasks.

The tasks of the final (persuasion) phase include such things as organizing the proposal into an outline, preparing a PowerPoint presentation, assigning roles in the presentation, practicing the presentation, giving the presentation, fielding questions, and preparing a final written report. The four reasons that will be offered to the company decision maker(s) to induce him/her/them to implement the proposal are four:
1. The proposal addresses the cause(s) of the problem.
2. The proposal will work.
3. The proposal is feasible.
4. The proposal will also help to solve other problems the company is facing.[46]

---

[46] For a fuller discussion of how to use these reasons to persuade decision makers to enact the proposal, see Lindsay, *Persuasion*, pp. 83-122.

So, what tasks will be required to prove these four reasons? Now, the students have completed the first step in a planning report. They have listed all of their tasks for the assignment. The rest of the planning report will proceed much faster.

Step two involves prioritizing the tasks. Much of this work has already been accomplished. Phase one (problem identification) is the first priority. The students cannot prove that a problem exists (phase two) until they have identified the problem they will solve. They cannot perform causal analysis (phase three) to determine, for example, *immediate* causes until they have shown in phase two when the trend began. They cannot analyze solutions (phase four) to determine which solutions address the cause of the problem until they have determined in phase three what that cause actually is. They cannot develop their proposal (phase five), which meets the three criteria, until they have analyzed all solutions in phase four according to the criteria. They cannot persuade their audience (phase six) that the audience should enact their proposal until they have actually developed the proposal in phase five. Hence, quite a bit of prioritizing has already occurred. There is, however, more prioritizing to do. The students may not have time to accomplish every single task they have listed. So, they must prioritize in each phase. What tasks are most likely to help us accomplish our goal at each phase? What tasks are next? What tasks are least important?

Step three involves allocating time for each task. Very simply, how long will it take to accomplish each thing? Step four is scheduling. How much time do we have to complete the assignment? Do we have minor deadlines along the way? Businesspersons use a chart to perform this scheduling. The chart is called a Gantt chart. Down the left-hand margin all of the tasks to be completed are listed in their order of completion. Across the top are listed the dates from the beginning to the final deadline. The chart then consists of shadings for each task, corresponding to the horizontal line from the task to the vertical lines of the dates. If it will take one week to complete a specific task, the shaded line begins at the first of the week it is scheduled and extends to the end of that week.

The planning report typically consists of a cover letter or memo explaining in paragraph form the phases students will go through in accomplishing their assignment plus appendixes such as the Gantt chart, a References page, and a problem-solving matrix that lists questions that need to be answered, possible sources for the answers, methods to be used in obtaining the answers, and personnel who are charged with finding those answers. However, essentially the planning report is nothing more than an elaboration of the basic relief valve for chrono stress--the four basic steps of time management:

1) List all tasks to be performed,
2) Prioritize these tasks,
3) Allocate time for performing each task, and
4) Schedule the tasks, beginning with the first priorities.

Chapter 6

Competence Stress

Johnny Carson contends that show business contains an abundance of shy performers, including himself. This is amazing to many. The concept of a shy communicator is so puzzling. Most believe that anyone who communicates for a living feels quite competent in all areas of communication. As a professor of communication, I can assure you that is not the case. But one category of communicators appears to experience no competence stress whatsoever about communicating. Newborn babies are poor communicators, but that does not stop them from blasting out their messages. They are not yet feeling competence stress. They may feel corporal stress--from a wet or dirty diaper, or from hunger, or from a headache or an earache, or from a skin irritation, or from a gas bubble, or from an upset stomach, etc. But WHICH corporal stress is it? They cannot tell you. You must guess. Yet you definitely know they are trying to communicate something. Babies are poor communicators--but they have no competence stress about communicating.

The fact that someone feels no competence stress about communicating does not mean that s/he is actually competent.   I know one individual who loves to communicate--but he is terribly boring.   He feels no competence stress whatsoever, but he is an incompetent communicator. Conversely, the fact that someone feels competence stress about communicating does not mean that s/he is actually incompetent. Johnny Carson and others are excellent communicators--but they have a great deal of competence stress about certain forms of communication. Johnny sees himself as inept at party conversation, for example. Why? Johnny describes some incidents of his youth.   He says he was the class clown.   He received immediate signs of approval when he clowned around--laughter.   The laughter made him feel more secure, more confident, and more competent at that type of communication. Yet, at parties, chitchat does not afford the opportunity for an audience to extend such approval. Certainly, there will be laughter if he tells a joke, but much of what happens in conversations is just routine give and take.   There may be no immediate signs of approval.   For a person who experiences communication competence stress, this may be unacceptable. Therefore, Johnny describes himself as shy.   Shyness is simply a term used to indicate that a person feels some communication competence stress. Hence, that person typically avoids those communication situations in which the stress is felt.

A woman I know experiences communication competence stress in a way exactly opposite to Carson's. She loves to chitchat. She can converse for hours on end with anyone under the sun, but she is extremely frightened at the prospect of communicating in public. In the same way Carson dominated his television show she dominates a conversation. Most people would never consider her to be shy, but she is--in public communication situations. She feels tremendous competence stress there.   Hence, she avoids public communication situations.   Many people do. According to public speaking experts, Jo Sprague and Douglas Stuart:

> Whether they call it stage fright, shyness, or speech anxiety, all speakers feel some fear. One out of five people experiences rather serious fear, enough to adversely affect performance. One out of 20 people suffers such serious fear of speaking that he or she is essentially unable to get through a public speech.[47]

I personally am closer to the Johnny Carson paradigm when it comes to communication competence stress. As a child, I found it virtually impossible to carry on conversations with people I did not know well. I was very shy. I avoided those situations. Then, as a young man I began to receive some reassurance

---

[47] Jo Sprague and Douglas Stuart, *The Speaker's Handbook*, 5th ed. (Fort Worth: Harcourt College Publishers, 2000), p. 73.

regarding my communication performance skills. I was told I had a nice speaking voice. I devoted myself to studying the art of public speaking. As a high school junior, I took a course in speech communication at a nearby college. I participated in local, state, national, and international speech contests and won. I still felt so much competence stress about coming up with things to say on the spur of the moment, I had to write out all of my speeches, memorize them thoroughly, and spend hours rehearsing them before I gave the speeches. During my speeches, people paid attention to me. That was an immediate sign of approval. They laughed at my jokes. Tears came to their eyes as I related emotional stories. They clapped at my character impersonations. They complimented me afterward. Today I feel very little competence stress when speaking in public because I have received approval, studied the art, and practiced the skill so much it has become what Sprague and Stuart call "unconscious competence." This is the final stage in the removal of competence stress. In the automobile-driving example from Chapter One, it is the stage in which the driver feels so comfortable; s/he can drive with one hand, eat a taco with the other, and carry on a cell phone conversation through his/her headset. According to Sprague and Stuart, this stage can be defined in this way:

> Now a person has integrated the learned skills well enough that she or he need not devote conscious attention to maintaining competence--it comes naturally. The skill becomes relatively effortless, and maybe even fun.
>
> At this level as a speaker you can do more than just talk; you have the freedom to pay attention to audience response and make spontaneous adjustments to enhance the quality of the communication.[48]

In other words, shyness (communication competence stress) can certainly be overcome. When my children were small they would never cross their teachers; they didn't have the nerve to order their own food at a McDonald's Restaurant; they felt enormous competence stress (shyness) in other communication situations. Shyness may even be a hereditary trait. My older daughter is a perfect example. She inherited her mother's shy quietness. As a child, she exhibited her shyness more as a reluctance to say anything than as a tendency to hide from people. She seemed to enjoy social interaction, but often simply listened to others talk for lack of anything to add. She entered Kindergarten a year earlier than her age dictated as a result of IQ and psychological testing at Purdue University. The only

---

[48] Sprague and Stuart, p. 14.

difficulty that the Purdue examiner experienced with my daughter's testing was her quietness. She often simply refused to answer the questions posed to her even though the examiner could tell she knew the answer. Would you believe she grew up to be a university professor? She feels very little competence stress when communicating with her students.

Not only can individuals overcome competence stress, they can also acquire it. My older son appeared to be a born verbal communicator. He was just a few months old when he learned to communicate with tremendous passion his love for McDonalds French fries and Coke. At the tender age of eight months he could identify the McDonalds golden arches blocks ahead of us when riding in the car. He would squeal with joy: "Non-nies!" (his word for McDonalds). He repeated the term with increasing excitement the closer we came to the giant arches. If we stopped, he would give his order: "Bop, Fwies!" (his words for Pop and Fries). If we passed by a McDonalds without stopping, his voice would trail off from excited expectation to disappointed resignation in a crescendo followed by attenuation of "Nonnies . . . NONNIES . . . NONNIES . . . N O N N I E S . . . NONNIES . . . NONNIES . . . Nonnies . . ." that would finally be curtailed by his waving at the golden arches dejectedly and saying, "Bye Bye, Nonnies!" He amazed his family and friends by learning such things as the German song, "Guten Abend, Gut Nacht," and the Greek and Hebrew alphabets at the age of one. He could virtually memorize and quote his children's story books with little effort, and seemed to be able to render all of the lines of entire movies like "Star Wars" and "Empire Strikes Back" after only one or two viewings. He demanded the right to put on performances for any visitors we received at home (known or unknown). Yet, for some reason, he became shy for several years during elementary school. He overcame his communication competence stress in college. My two younger children seemed very confident and outgoing throughout elementary school. Then in high school my son, and in college my daughter, became shy for a while. After a year or two, my younger son began to receive many signs of approval for his acting ability in high school and community plays. He overcame his competence stress in a sense similar to the Johnny Carson paradigm. He worked as a cast member in Walt Disney World. He gave guided tours through the safari in Animal Kingdom. My younger daughter also turned to performance as a musical performer.

The point of my survey of Johnny Carson's, my family's, and others' communication histories is to dispel several myths about competence stress. Those myths are the following:

- *Competence stress is a permanent condition.* Obviously, competence stress is greatly relieved when confidence is gained. As a parent, employer, or friend, you can relieve your children's, employees', and friends' stress by simply giving more signs of approval. As a person who experiences competence stress, you can relieve your stress by practice and preparation. As I devoted myself to learning the art of public speaking, devote yourself to learning whatever skill at which you are feeling incompetent. And practice! Driving a car without becoming tense and stressed out is simply a matter of driving and driving and driving until it becomes second nature to you--until you reach the stage called unconscious competence.

- *The absence of competence stress in someone indicates that s/he is competent.* Babies dispel this myth; they are neither competent nor stressed about it. The boring man who loved to talk was neither competent nor stressed about it.

- *The presence of competence stress in someone indicates that s/he is incompetent.* Johnny Carson and other shy performers dispel this myth; they are not incompetent but they do have competence stress. My children's individual bouts with shyness were no indication that they were incompetent communicators.

- *Competence stress does not affect all people.* Everyone feels competence stress about something. Michael Jordan may experience no competence stress about playing professional basketball, but playing professional baseball is another matter. Johnny Carson may experience no competence stress about performing on television, but carrying on a polite conversation is another matter.

- *It is easy to determine the cause of competence stress.* On the contrary, it is often difficult to figure out what is causing the stress. Is shyness hereditary or environmental? My older daughter and son exhibited vastly different patterns of developing and overcoming competence stress. We thought the cause in one instance was hereditary, but that was later dispelled. Perhaps the change of environment from a small town school system to a small city school system caused my older son to experience competence stress. Perhaps a broken collarbone, which caused my younger son to miss a year of athletic competition in middle school, affected his stress level. Perhaps the move to Indiana University for my younger daughter negatively impacted her competence stress level. Her older brother and sister had attended a small private college. Maybe a critical-though-possibly-absent-minded comment from someone produced the stress. Maybe they came in contact with others who appeared to them to be more competent. They may have lost confidence as a result of the comparison. Perhaps, like Johnny Carson, we all need signs of approval and if they are not immediately forthcoming at any given point, our response is to experience increased competence stress. Whatever the cause(s), the relief valves are basically three: practice, preparation, and receiving signs of approval from others.

I have emphasized competence stress pertaining primarily to the art of communication in my examples thus far. This was for demonstration purposes only. People experience competence stress in virtually any activity in which humans engage. The word "competence" contains the word "compete." Every athletic contest in which individuals compete affords opportunity to experience competence stress. Would you like to be at the foul line shooting free throws when the time on the basketball game clock has run out and your team is trailing by a point? If not, you would have competence stress in that situation. If your answer is "yes," you have probably reached the "unconscious competence" level in basketball. People also compete in the arena of looks. If someone feels s/he does not look as attractive as others in some respect (clothing, make-up, hairstyle, physique, height, weight, lips, nose, legs, chin, fingernails, cheeks, ears, etc.), competence stress may result. Even skills such as parenting, kissing, cooking, sewing, singing, painting, money managing, negotiating deals (such as car buying), car and home maintenance, etc. have a "competing" or competence stress element. If everyone in the world were just as inept as you at the computer, for example, you would have no competence stress about doing computer work. There would be no one competing with you in that skill. But, begin to talk to a bona fide high tech computer nerd and, even if you felt fairly competent in the past, you immediately begin to feel some competence stress in that area.

Students experience competence stress when taking quizzes and exams. I hate giving final exams to students. The pressure is too great. Why should a final grade in a course that has taken an entire semester to complete come down to the results of a single exam? I prefer to spread the quizzes and exams out over the length of the semester. I remember taking my first preliminary exam in my first Ph.D. program. What stress! I had taken years of courses and received A's in virtually all of them. Yet, if I did not do well on this exam, it could all be lost. Even here, though, practice and preparation were great relief valves. By the time I took my final oral examination for my Ph.D. at Purdue University, I was so confident in my knowledge and ability to take the exam, it was (almost) a piece of cake. Lawyers must pass the bar exam. CPAs have to pass exams to be certified. The only relief valves available to exam takers are practice and preparation.

Quite similar to the exam pressure of students, lawyers, and CPAs is the competence stress salespersons feel when giving extremely high potential income-producing sales proposal presentations. There is little stress involved in retailing items with low mark-up. But, when the pending pay-off could be thousands or millions of dollars, you feel like a doctoral student standing for his or her final oral exam. This sales competence stress is also similar to competence stress in athletic contests. Consider the poem, Casey at the Bat. The result ("mighty Casey had struck out") is the basis of fear associated with competence stress. If everything is on the line and it all comes down to your competence, you may feel competence stress. Yet, just as there are athletes who appear to have "ice water in their veins," there are practitioners of virtually every skill who feel so competent they want the opportunity to perform under the most demanding circumstances. Sometimes this is due to a desire for eustress. Sometimes it is due to the fact that there is little competence stress. Michael Jordan does indeed feel more competent at taking the final shot in a close game than anyone else. Kenneth Burke states it this way: "People who are expert at solving puzzles will prefer hard ones."[49]

Most new employees experience competence stress. A young college graduate called me after one week on the job. He had received a very nice salary offer but was devastated by the amount of information he did not seem to know as he began his first job after college. He feared he would lose his position due to his incompetence. I told him to stop worrying. His company was the entity taking the risk here. Even if he lost his job he would have received excellent income for whatever number of weeks or months he worked. He was not making any money before he took the job. So, he could not lose by working for this company for as long as they wanted him. Yet, the company would lose all of the money it had invested in him if he were fired. I told him to consider the whole thing a learning experience. Even if he lost the job he would surely know more about his career when he started to look for his next job. So, he stuck it out. He was a fast learner. He received raise after raise after raise, as he grew more and more valuable to the company. After six years he was still working for the same company and climbing the corporate ladder. Practice was all it took to relieve his competence stress. He just needed to keep doing things until the stress subsided.

---

[49] Burke, *Language*, 298.

Kenneth Burke's definition of human includes the clause: "*goaded by the spirit of hierarchy (or moved by a sense of order)*."[50] Burke is claiming that humans constantly compete to attain higher levels in various hierarchies. The corporate ladder is one such hierarchy. Another hierarchy is the family hierarchy. We call the competition among children in this hierarchy "sibling rivalry." Politics is a hierarchy. Politicians may experience competence stress. Will I be able to raise enough funds to compete successfully with my opponent? Will I be able to debate successfully? Will I be able to come up with campaign strategies as successful as my opponents? Do I look good on TV? Do I sound good on the radio? Am I too short? Too fat? Too old? Do I know enough about the issues? Even religions have hierarchies and their inherent competence stresses. The word "hierarchy" actually comes from religion. The element "-archy" means "first" as in the study of "archaeology"--the study of "first" things. The element "hier-" is from the Greek word *hieros*, meaning, "priest." There are various "priestly" levels in the Roman Catholic Church, for example: Pope, Cardinal, Archbishop, Bishop, Priest, etc. I would imagine that those who wish to compete for the "first" or higher positions among priests experience competence stress as they compete.

Burke's point, however, is that humans create all kinds of symbolic hierarchies-- from the best tobacco spitter in Tennessee to the best practitioner of speaking the English language in Britain to the best looking hand model in Hollywood. There are millions, nay trillions, of hierarchies. If you (or someone you know) experience too much competence stress because you have low self-esteem, I suggest that you are in the wrong hierarchy. You are competing with others who may have more expertise or ability or gifts in that specific skill or attribute. Whenever you compete with others you may experience competence stress. If you are clearly at the top of a hierarchy, you experience very little competence stress. Just choose (or invent) a hierarchy in which you are at the top. Everybody has one. My uncle Joe was born with Down's Syndrome. There were many things he could not do. But one thing he loved to do was greet people on Sunday morning as they arrived at church. He smiled and shook hands with everyone. He would have been pretty low on most symbolic hierarchies, but there was one hierarchy in which my uncle Joe was at the very top. He was probably the best greeter with Down's Syndrome in the independent Christian Churches and Churches of Christ fellowship in the entire world. At his death, a tribute to him was published in one

---

[50] Burke, *Language*, 15-16.

of the most widely distributed Christian magazines in America.[51]   I offer salespersons a quite successful technique for placing themselves at the top of a hierarchy in my book *Persuasion, Proposals, and Public Speaking*. I borrowed the technique from John Savage and adapted it to reflect Burkean insights.[52] Hence, in addition to the three relief valves mentioned earlier, a fourth relief valve for competence stress may be to choose a new hierarchy. Don't compete on someone else's turf. Compete on your own! You will encounter less stress.

There are definitely times, however, in which you are highly motivated to compete in a skill at which your level of competence is low. I am an educator. I believe in preparation.   I believe that individuals should seek to improve themselves in areas in which they are incompetent if there is a good reason to do so. Preparation is the relief valve needed in such instances. Go to school. Study on-line. Read voraciously. Attend seminars. Do research. In short, prepare. You will find that you go through essentially four stages in developing competence. Sprague and Stuart list these stages. The fourth stage--unconscious competence-- has been previously mentioned.

The first stage is unconscious incompetence. "In this stage a person is not aware that he or she is making errors in some area, and may even be unaware that there is a skill to be learned."[53]   Babies are at this stage in communication skills. You could also call this stage naïveté. The boring man I mentioned might not even realize that conversation is a skill that can be learned. When I coached elementary school basketball players, it was easy to see that they were at this stage with regard to basketball skills. There is very little competence stress here. This explains why some people may be stress-free at first, then develop competence stress later.

The second stage is conscious incompetence. "A person at this stage has made the realization that she or he is doing something ineptly and that there is room for improvement.   In many cases this awareness creates anxiety, which actually increases incompetence."[54] How do you learn that you are incompetent? Perhaps someone mentioned it to you or perhaps you observed the fact as you compared yourself to others. This is the increasing stage in competence stress, but it is not necessarily a bad thing. It has been said that nobody likes criticism. While that may be true, most of us can benefit from it. We do not like it partly because it

---

[51] Karen Norheim, "A Heart for God," *The Lookout* (January 29, 1989). 8.

[52] Lindsay, *Persuasion*, 79-80.

[53] Sprague and Stuart, 14.

[54] Sprague and Stuart, 14.

increases our competence stress. As I offer criticism (to my students and others), I try to keep in mind seven suggestions made by Joseph DeVito[55]:

1. Say something positive. Our English word "criticize" comes from the Greek word *krinô*, meaning, "to judge."   Judges are not always negative. Sometimes judges declare people "not guilty." Tell the person you are critiquing what s/he is "not guilty" of. What is s/he doing right?

2. Be specific. Nothing is accomplished if you tell someone s/he is poor or s/he is good. What specifically is poor? What specifically is good? If I am incompetent when using a rod and reel to catch fish, do not just tell me I am a poor fisherman. Tell me specifically what I am doing wrong.

3. Be objective. Although it is probably impossible to entirely remove your own prejudices and opinions from your critique, try to. Be aware of your prejudices and ask yourself, "Are my prejudices affecting my critique?"

4. Limit criticism. Even Bobby Knight who is well known for his stinging criticism of his star basketball players goes lighter on his less-talented players. If his stars make mistakes, you can bet his less-talented players make many more. Yet, if he were to get on his less-talented players for every mistake they make, they would be totally devastated. His more talented players know that since they have less competence stress, the coach is adding some. Every critic should consider how much criticism may be beneficial and how much would actually be detrimental. As Sprague and Stuart point out, "In many cases this awareness creates anxiety, which actually increases incompetence."[56]

5. Be constructive. It is not enough for a critic to point out what someone is doing wrong. The critic should also explain how to improve. This is called constructive criticism. Be careful, however. The one offering the constructive criticism should normally have more expertise than the one being criticized. Otherwise, why should the one being criticized listen to the critic?

6. Focus on behavior. It is easy for critics to think they can read the mind of the one they are criticizing. Avoid such presumptions. Critics are not thought police. Only God knows what other people are thinking. But critics can judge observable behavior. Focus on critiquing that behavior.

---

[55] Joseph DeVito, *The Elements of Public Speaking,* 6th ed. (Longman Publishing Co.), 82-86.
[56] Sprague and Stuart, 14.

7. Own your own criticism. If you follow DeVito's first six suggestions, you should have no problem accepting personal responsibility for your criticism. You have critiqued fairly, constructively, objectively, with positive comments as well as negative ones. The one form of critique I personally detest is the anonymous attack. The goal of this criticism is not improvement; it is usually the venting of frustrations or even hatred. If everyone owns his or her own criticism, such hate communication will be less likely to occur.

Nevertheless, as you enter the stage of conscious incompetence, you should expect your competence stress levels to rise. You are discovering new areas in which you are not fully competent. This is somewhat stressful, but handle the stress with humility. You know that you need improvement in these areas and you are determined to prepare and practice until you improve.

The third stage of developing competence is conscious competence. According to Sprague and Stuart, "In this stage a person has taken a skill in which she or he feels incompetent, has improved, and then devotes a portion of consciousness to performing it competently. . . . [I]f a person perseveres, the awkwardness of the new behavior diminishes and the need for self-monitoring lessens."[57] As you enter the stage of conscious competence, you should expect your competence stress levels to decrease. The stress is not completely absent, but you know you can be competent so long as you devote some attention to what you are doing. With reference to my driving example in Chapter 1, this is the stage at which you are a competent driver, but you still must keep your hands at the ten o'clock and two o'clock position. You must always keep your eyes on the road. You will not allow other activities such as eating, talking on the phone, or perhaps even conversing or listening to the radio to distract you. You are competent, but you are still experiencing competence stress.

The fourth and final stage of developing competence has been discussed previously--unconscious competence. You are good at what you do and you know it. You feel very little competence stress whatsoever.

Recapping, then, the primary relief valves for competence stress are practice, preparation, and receiving signs of approval. Additionally, it is helpful to choose your own hierarchies. Don't compete on another person's turf; compete on your own. Finally, if you find that you should enter arenas in which you are

---

[57] Sprague and Stuart, 14.

incompetent, prepare diligently. Move through the four stages of developing competence, but do so with humility. You will be amazed how much competence stress is relieved if you are humble. Tell people you are still learning that skill. Ask them to be patient with you and they probably will be. Their expectations as well as your own will be lessened for the moment and the competence stress will be decreased.

Chapter 7

Confusion Stress

Confusion stress is the stress you feel when you are lost. You felt it when you were a child and became separated from your parents at a large store. You felt it when you were traveling and suddenly realized you did not recognize any of the road signs or your surroundings. You felt it when you were taking an educational course that was way over your head. You felt it when you were attempting to win a court case and the shyster lawyer who opposed you confused every issue.[58] You felt it when your religious faith was challenged to its sheer underpinnings. Kenneth Burke describes this last type of confusion stress as it reaches distress levels: If a superior debater systematically questions everything one says and is persistent in his/her questions, "piling one atop another, and stopping just short of the point where the person questioned is about to explode with anger, . . . the

---

[58] Burke, *Language*, 414, suggests that Marshall McLuhan used the same resources of debate when his philosophy of media was challenged: "For he uses question periods not as opportunities to make his position more precise, but rather as challenges that he must deflect and *confuse* to the best of his ability." (Emphasis mine)

thought of ultimate questions makes one feel the underpinnings of one's beliefs melt away."[59]

I grew up on a farm in Illinois. Surrounding my home were huge fields of grain--wheat, soybeans, and corn. As a child, I typically stayed out of the wheat fields except when the plants were young; then, I might ride my Shetland pony through the fields. I did not care to spend time in the bean fields because most of my time spent there was work time--walking the rows of beans and cutting out weeds by hand. The cornfields, on the other hand, were always fascinating. The cornstalks grew to a height of at least six or seven feet. Except for the blades on the stalks, which tended to scratch and cut me as I walked through them, I found cornfields to be a great place to play. I could hide in the cornfields. Once I moved a few rows into the field, no one could see me. One day when I was about seven years of age, I was playing in the cornfield, moving from hideout to hideout when I realized that I had not been in this specific location before. I looked up but the sun, clouds, and sky I saw filtering through the tops of the cornstalks (more than twice my height) gave me no indication in which direction to travel. I shouted loudly for help, but no one heard me and I could hear no one. I began to walk through the stalks of corn in what I thought was a homeward direction, but I did not reach home. I changed directions and walked further, but still I did not arrive at my home. I was lost.

The confusion stress involved in being lost comes from not knowing. I did not know where I was. I did not know in what direction I should move. I did not know if anyone had noticed I was missing. I wept as I continued to push through the slicing green blades of corn, but it was useless to make a great effort of crying--no one was there to comfort me. Yet, as discussed in Chapter 2, weeping itself is a universal relief valve. It probably helped me control my stress. Even as a child, I began to formulate a solution to my dilemma. Although the fields were huge, covering perhaps one hundred acres per field, there were edges of the fields in every direction. I knew most of the edges. Roads, levees, fences, ponds, and/or the Sangamon River surrounded most fields. I would walk until I reached some edge. Then I would try to determine at what edge I had arrived and move in the proper direction. It was a long day for a child. I believe that I spent nearly five hours wandering through the cornfield. Eventually, I came to a field road. It looked familiar to me. I had traveled the field roads with my father in trucks and cars and on tractors. This road was on the end of the cornfield opposite my home.

---

[59] Burke, *Language*, 459.

I began to walk the road in a direction I thought would take me home. After a few minutes of walking, I saw a car coming. My parents had been driving around frantically for hours searching for me. Their stress seemed to me to be greater than my own. Apparently, they had felt confusion stress themselves--they had lost their child.

The confusion stresses my wife and I have encountered together have not been limited to this continent. Our Baltic cruise to Germany, Denmark, Sweden, Finland, and the Soviet Union resulted in confusion stress. The site was Leningrad, USSR, at the time (now, St. Petersburg, Russia). We had just finished touring a former cathedral, which the Communist government had converted into a museum of atheism, when a man approached us. I had been speaking to my wife in the German language in a joking effort to get into the "foreigner" character (since we could not understand any of the language that was being spoken all around us).

"Sprechen Sie Deutch (Do you speak German)?" the man asked me.

"Ein bischen (A little)." I replied.

He tried a new language: "Then, do you speak English?"

"Yes," I said.

He proceeded to offer to trade Russian rubles to us for American dollars.

"Sorry." (We weren't carrying much cash.)

He noticed a gold chain necklace, which my wife was wearing, and a watch on my wrist that had already begun to malfunction for me. He offered us about three months Russian income in rubles to us for these two items worth only about twenty-five dollars totally.

We agreed.

Later we discovered that it was illegal to take rubles out of the country. We rushed to spend the Soviet currency. The Soviet citizens watched in awe as we spent the money so freely. We walked the streets of Leningrad searching for stores in which to spend the money. We finally exhausted our supply of rubles and were loaded with packages, trying to find our way back to the ship. We had a map of Leningrad, but the street signs were not written in English, of course. Although the man who traded us rubles for our jewelry spoke English, no one else we encountered spoke anything but Russian. We could not ask anyone for directions. The street signs were not even written using the English alphabet. The signs were written in Russian. Neither of us knew Russian, but somehow I recognized many of the letters. They looked like Greek letters to me and I knew Greek. I later found out that Russian uses the Cyrillic alphabet (reputed to have been invented by St. Cyril in a.d. 869). Apparently Cyril had used the Greek alphabet as a prototype for inventing the Russian alphabet. I was able to make out enough letters to read the street signs and find our way back to the ship.

Then, as we were returning to our ship, the realization hit us that we would be scrutinized by the border guards before we were allowed to return to the ship. These guards were not like agents on the U.S.-Canadian border. We had to undergo their extreme scrutiny multiple times as we entered and left the ship. The guards always dwelt awhile gazing at my wife. Of course, I enjoyed gazing at her too, but theirs was not the same purpose. Her passport picture had been taken when she had long, straight, blonde hair. Before we left on the cruise she had gotten a short cut and a perm. The hair stylist had misunderstood which style she wanted and I teased her throughout the trip telling her that her hair looked like Harpo Marx's. She did look like a woman completely different from the one pictured in her passport.

I had already experienced brief scary dreams of losing our passports or the "travel passes" which the Soviets substituted for the passports, thus stranding us in Russia forever. Being in a strange country--with a language we did not know, a legal system we did not know but feared, and specific rules even regarding how one exchanges money, with which we were not familiar--also gave us confusion stress. Now we were approaching the guards near the ship. We saw some notices to the effect that all currency exchanges had to be made at the official currency exchange offices. Had we committed a misdemeanor by trading jewelry for rubles? A felony? We honestly did not know. Should we confess to the Soviet authorities what had transpired? Would they understand? Would it cause an international incident? We were confused and the stress that the confusion produced was overwhelming. My wife just knew that we would be thrown into a Soviet prison. As the stone-faced guards examined our passports and checked every facial detail against our pictures, they didn't bother to question us regarding the packages that we were carrying. We reached the safety of the Norwegian-registered ship without incident. Now the stress was relieved. We were in safer confines. People here spoke English and we knew what was expected of us.

Traveling to other lands is capable of producing great confusion stress. You often do not know the language or culture. Therefore, you are confused. We had less stress on the cruise ship than we would have experienced if we had traveled to Scandinavia by ourselves. We had tour guides and they spoke our language. I found also that, although their first languages were Danish, Swedish, Finnish, Russian, German, and Dutch, many Europeans had taken the time to learn English as a second language. We could exercise one major relief valve for confusion stress: Ask questions! Later, when my wife and I returned to Germany for a week by ourselves, we were less stressed than we might have been since I had studied German for a few years. I could ask questions in German and sometimes even understand the answers I received. Even when we visited Paris by ourselves we found Parisians who spoke English and the alphabet in most of these European countries is at least the same alphabet we use in English. We could read street signs.

It is not necessary, however, to travel to foreign countries to experience confusion stress. Another trip, which I earned in insurance sales, took us to Maui, Hawaii (in the good old USA). We also possessed a certificate worth two days free lodging near Waikiki Beach in Honolulu. We had received this certificate for touring a condo time-sharing presentation. I'm sure they really did not expect people to take a trip to Honolulu and only use the two days of free lodging, but we did. We arranged our itinerary accordingly. On the morning on which we were scheduled to fly out of Indianapolis, I had misplaced our Honolulu reservations. Confusion stress--I had "lost" my reservations. We searched frantically and unsuccessfully for an hour, then realized that we might miss our plane if we searched longer. We then drove desperately toward Indianapolis. The realization set in after a few minutes of driving that there was no way on earth that we could arrive in time to catch our flight. We now had confusion stress compounded by chrono stress.

Plan B! West Lafayette, Indiana, is on the interstate highway between Indianapolis and Chicago. We turned around and started driving even more desperately in the opposite direction. Our tickets called for making connections at O'Hare Airport in Chicago. We calculated that if we traveled fifteen miles per hour over the speed limit we could arrive five to ten minutes before our flight would leave Chicago. It was almost working, too . . . until we neared the Chicago area. I looked in my mirror and saw red flashing lights coming up behind me! Now our confusion stress and chrono stress was compounded by conscientious stress! I knew I had violated the law.

I pulled over. The police officer escorted me back to his squad car. I quickly explained our dilemma and said, "If you have any mercy in your heart, whatsoever, I would sincerely appreciate it today." The officer did have mercy. He wrote me a warning ticket. As we will see in the next chapter, forgiveness is a relief valve for conscientious stress. By writing me a warning, the officer in effect forgave me of my infraction. You could say he also relieved some of our confusion stress. I used the relief valve--ask questions. I asked and he gave me directions for the quickest route to O'Hare. We arrived at the terminal three minutes before the scheduled take-off. I dropped my wife and our baggage off at the curb with a baggage carrier who rushed her to the gate. I sped to the nearest (and most expensive) parking area and ran to the gate. My wife had arrived in time. The plane was waiting for me to arrive before it departed. We hustled aboard the plane (after tipping the baggage carrier generously) and flew off to Hawaii! Upon our arrival in Maui, I phoned our condo in Honolulu explaining that I had lost our reservations. No problem. They would honor our free lodging anyway! Our confusion stress was once again relieved by simply asking questions.

However, the stress could have been better relieved if I had exercised a different relief valve for confusion stress--organization. Like the person I mentioned in Chapter One who had lost his car keys, I should have kept everything organized so I knew where all of my documents for the trip to Maui were. Those who are prone to misplace plane tickets, assignments at work or school, their glasses, important bills, diamond rings, wallets, purses, or play tickets would do well to invest some effort in organization.

I mentioned my thirty-six-slot organizer in Chapter One, but there are many possible organizational systems available. Use wall hooks for keys and tools. Use racks for shoes and sports equipment. Use closet organizers, drawers for clothing and smaller drawers and cabinets for nails, screws, bolts, etc. Use filing cabinets for tax papers, product warranty information, and projects (such as trips to Maui) you are working on. Organize your bookshelves so you can find the necessary texts. Can you imagine how much confusion stress you would feel if, as a college student or researcher, you went to library only to find shelf after shelf of books with no organizational system? Then, make your own home or office as organized as the library. (Well, maybe that much organization is asking too much!) The point is, if you are experiencing frequent confusion stress due to the fact that you have lost something again, your best relief valve is organization.

Businesspersons should certainly understand the need for organization. We even call most businesses "business organizations." I teach courses in organizational communication. It makes a difference in our style of communication whether you are doing upward communication (with a superior in your business) or downward communication (with an inferior) or horizontal communication (with an equal). It makes a difference whether you are doing internal communication (communication within your company) or external communication (communication with a client or someone else outside the organization). Certainly, you need to keep records. How else will you know if procedures you are following work? How will you know if you are using the best form of advertising? How will you respond to the IRS if they schedule an audit? You need to keep records.

Sometimes, despite your best efforts at organization, you still face confusion stress. Our second free trip to Hawaii came a few years later, with a different insurance company, but the confusion stress was equally severe. Our hometown of West Lafayette, Indiana, looked like Baghdad on the first night of the bombing raids in Operation Desert Storm--the war to liberate Kuwait. Frequent flashes of light from explosions lit a steadily darkening town. The streets resembled a war zone as downed power lines crossed many of the roadways. The explosions were power transformers exploding all over the city. Tree limbs snapped and came crashing to the ground. Whatever limbs and power lines remained above the ground level shimmered and glistened with inches of icy coating.

My wife and I were afraid to actually enter the city of West Lafayette for fear of being hit or trapped by falling trees and power lines but we had some last minute details for our trip to Kauai to handle. Our own electricity at home had been knocked out an hour before as we saw the power line outside our house sag, snap, and fall to the ground. I needed to find some place with electricity. My wife and I were leaving on our flight to Kauai, Hawaii, early the next morning and I had not yet put a new message on my answering machine advising my callers that I would be out of town for a week. We stopped at a Wendy's restaurant that still had lights. I carried my answering machine into the restroom and searched for an electrical outlet. None could be found. I drove to the Post Office and thought that I spotted an outlet hidden behind a row of temporary Post Office Boxes. When I squeezed behind them I found no suitable outlet.

We then remembered our church building to which we had keys. Though the church was just outside the city limits, it still had electricity. In fact, it was located just a few hundred yards from several transformers. I recorded my new message, the kids did some homework in the light, and we all used the restrooms before heading home by back roads. We feared the exploding transformers in town and were somewhat worried by our proximity to the several transformers near the church.

My wife had not yet finished our laundry but the dirty clothes could be packed and taken to Hawaii for laundering there. The darkness enveloped our home and small subdivision. We went to bed since we needed to get up and leave in the middle of the night. The alarm clock sounded soon. Our house was still without electricity--no showers for the long flight to Hawaii. And the temperature inside the house was steadily declining. Worse yet, the ice storm that had hit us turned out to be bad. We had been organized. My wife's sister and my parents had planned to take turns staying with the kids while we were gone. However, the roads between our house and my sister-in-law's were impassable. We heard on the portable radio that the police had closed all roads in her county. And we knew that the kids would need heat, running water, and bathrooms while we were away. We had also stocked the refrigerator with food for our kids and their supervisors, in anticipation of our absence. We had been organized. Now, the electricity was off and the food was in danger of spoiling.

The confusion stress was becoming severe. In those wee hours of the morning we concocted a new strategy. We picked up the phone and dialed my wife's brother and his wife. They hadn't even known that we were going to Hawaii for the week but, in the crisis, they willingly offered to help out until my sister-in-law could get through. (That turned out to be a long "until." The power blackout lasted nearly a week.) The stresses that my brother-in-law, sister-in-law, and the kids endured that week were monumental. When we touched down in Kauai, I went to the phone and called our house. There was no answer. I then called my brother-in-law's number and his wife answered. The kids were with them.

Everyone's electricity was out but my brother-in-law had a wood fireplace which took a little of the chill out of the air and, since he lived within the city limits, he had city water supplied by a water tower. The water flowed by force of gravity, not electric pump. Still, he and his wife worried that somehow they might lose their water supply so his wife instructed her son to fill the bathtub upstairs. Several minutes later they noticed that their ceiling was leaking. My nephew had forgotten to shut off the water.

Meanwhile, we had seen no ocean at all from the plane, which carried us for hours over the Pacific. Clouds blanketed the entire earth, for all we knew. We thought we would eventually fly out of all of the clouds but we didn't. Almost our entire week in Kauai was spent under cloud cover with strong trade winds blowing in cool air. The surf was too rough and cold to enjoy the beach outside our room. Even our "whaling" tour in which we spotted one brave whale in the ocean waters was marred by the extremely high waves and the extremely deep swells which our tour boat had to navigate as we went whale watching. We would at times dip into the depths of a swell and look up to see the height of a wave towering twenty feet above us. If we had known in advance what the full impact of the storm would be and the kind of weather we would be encountering in Hawaii we would have skipped this trip. Yet, somehow everyone survived. The kids didn't even have school for the week due to the effects of the ice storm. As with my childhood adventure in the cornfield, some confusion stress simply has to be endured as we try to be resourceful. We adjust our strategies with each new confusing development.

Handling crises falls under this relief valve for dealing with confusion stress--be resourceful; be willing to adjust your strategy. As a financial consultant, I have recognized the need for personal life insurance. I know that my family could face a financial crisis on any given day. I could be killed in an automobile accident, for instance. I used to commute (sometimes hours) to the colleges and universities at which I taught. One day, I was driving an hour on the interstate to teach at a university in Indianapolis. It was early morning, dark, and very rainy. My wife had warned me not to use the cruise control on our van during the rain but I disregarded her advice. Suddenly, I noticed my van hydroplaning and the rear of the van starting to sway back and forth. I tried to remember quickly how to shut off the cruise control without using the brakes. I knew that, if I touched the brakes, the cruise control would shut off but I might lose whatever tire traction I still had. The controls for shutting off the cruise were located on the steering wheel somewhere, but I could not afford to take my eyes from the road to locate them, nor my hands from the steering wheel to touch the button. I could not think of another way to shut off the cruise control and I was starting to lose control of the van. I chose the foot option. I touched the brakes and my van began to spin uncontrollably. Fortunately, it continued to move in the same direction as the highway, albeit spinning furiously in circles. Finally, it came to rest on the pavement of the highway, but sideways, blocking both lanes of traffic traveling my direction.

I was relieved that I had not slid into a ditch and turned over several times. I was relieved that I had not hit another vehicle, road sign, or overpass column. I stepped out of the van to survey the situation momentarily. But, then I noticed that the oncoming traffic, however sparse at this early hour, was not slowing down as it approached me. Could they not see me since my van was turned sideways? Neither my headlights nor my taillights were visible to them. And two vehicles were approaching side by side--one of them a semi. I was now in full-blown crisis planning mode. I didn't know what to do--confusion stress. If I stayed in the median, one vehicle might swerve into the median to avoid my van and hit me. If I stayed in the van, they might hit the van, but at least I would have some protection against the collision. I chose the latter option, swiftly climbed in the van, fastened my seat belt and braced myself. The semi swerved onto the shoulder in front of me and avoided me. The other car swerved into the median behind me and avoided me. If I had remained in the median, I would have been killed. With that crisis past, I thought I would start the van and back it off on the shoulder before the next vehicles arrived. I turned the key in the ignition, but it would not start. Now what? I decided quickly. I turned the key to the "on" position, moved the gear shift level to neutral, opened my door, and pushed the van off the road and into the median just before the next few vehicles barreled through. I had lived. I had a few problems that remained to be solved, such as getting my car out of a median that was quite muddy due to the heavy rainfall, but the crisis had passed. My body had pumped adrenaline into my system and my stress level had been sky high. I had used the resourcefulness relief valve and had continuously adjusted my strategy to survive.

Sometimes, the crisis is not as short-lived as the event I just described. Every person who has been diagnosed with a potentially fatal disease knows what I mean. For some of my friends, relatives, and associates, the disease of crisis was cancer. For me, it was heart disease. I have never actually had heart disease, but I found out suddenly that I have the potential for it. As an insurance agent, I applied for additional coverage on myself from time to time. I always qualified for the best rate in the specific company to which I applied. I had, and continue to have, excellent blood pressure. My EKGs always come back perfectly normal. My heart sounds good and my cholesterol levels had (in the past) always been fine. I applied again and again to various companies and I always needed to undergo a thorough insurance physical in each process. Finally, I took a physical that did not come back with the same results as the earlier physicals. The insurance underwriters ordered a new blood test. Then, they sent me the message that I had been declined for coverage. Not rated up. A rating would have indicated that a problem was developing. But, I was declined. I knew what that meant: I stood a good chance of dying in the not-too-distant future. Otherwise, the company would have attempted to offer me SOME insurance at SOME price. But, there was no rating--nothing. I was declined. Although I knew he was not legally allowed to tell me, I asked the underwriter what the problem was. He said he could send the report to my physician, but he could not tell me specifically. Could he at least tell me what type of malady this was related to? Yes, it is related to heart disease.

Now, I was experiencing strong confusion stress. There were many things I did not know. What was my exact problem? Was there a cure? What was the prognosis? How long would I live? Would I even survive long enough to talk to a physician? Was I about to have a heart attack? The confusion stress was intense. I could not sleep at night. Although I joked about it, I was quite worried until I had an opportunity to speak to my physician. I invoked relief valve number one. I asked the questions of my physician. He told me my problem was that my bad cholesterol was skyrocketing and my good cholesterol was taking a steep dive. The ratios between the two were terrible. Was there a cure? He said, "Let's start with diet and exercise." I avoided every single food with cholesterol for two months. I exercised daily. I still couldn't get enough sleep due to confusion stress. I did not know if this would work. I went in for further tests. The problem was continuing and even worsening. Cholesterol is a problem because it is gooey substance that tends to cling to the walls of your arteries, restricting the flow of blood. Calcification then occurs where the cholesterol has been clinging, resulting in hardening of the arteries. If the blockage of the arteries results in the stoppage of blood flow to heart muscles, those heart muscles die for lack of oxygen. This is known as a heart attack.

"Is there anything else I can do?" I asked.

"Yes. Let's try Zocor," he responded. Zocor was one member of a new category of cholesterol-lowering drugs. It posed some concerns. Occasionally, it causes liver damage, but my liver was pretty healthy. I have never consumed alcoholic beverages, one of the liver's biggest enemies. My doctor told me we would try Zocor in small amounts and test me again in three months both to see if the cholesterol was decreasing and to see if any liver problems were developing. My confusion stress continued, but at least I was trying the resourcefulness relief valve. I was adjusting my strategies. After three months, the blood tests showed significant improvement. The Zocor was lowering my bad cholesterol, but not enough. The doctor doubled my dosage. That worked even better and there were still no signs of liver problems. However, the Zocor was not lowering my triglycerides. The doctor changed my prescription to Lipitor, another drug in the same cholesterol-lowering family. My triglycerides started coming down along with my bad cholesterol, but my good cholesterol had not been rising at all during this treatment. What would help that? Red wine. But I do not drink alcohol. My doctor told me he would not recommend that I start drinking just for this problem. There are too many risks in drinking alcohol. That was fine with me; I would not have started drinking anyway.

Then, I heard a radio news story that claimed that Welchs (the grape juice company) had been conducting studies to see if the medicinal benefit was in the grape or in the alcohol (red wine). Welchs found that they achieved the very same heart-smart results with purple grape juice that they achieved with red wine. I began a daily regimen of drinking Welchs purple grape juice every morning. What do you know? They were right. My good cholesterol has been rising dramatically. My cardiologist was so impressed with my progress, he said he planned to prescribe Welchs purple grape juice for his other patients in similar situations. My confusion stress pertaining to my heart problems is virtually gone. I worked through it by asking questions, being resourceful, and continually adjusting my strategy.

Others who have faced illness-related crises have not been so lucky. Not everyone survives a life-threatening health condition. Yet, even though someone does not know if s/he will survive, confusion stress can be reduced by at least having a plan, asking questions, being resourceful, and being willing to continually adjust strategies. There is at times even a relief that comes when you are told the strategies have failed; you are going to die. At least, you now KNOW. You are no longer confused about the future. And, you know that you have tried your very best to win the battle. Clearly, religion is also a relief valve in such situations. Belief in the world to come is a comfort to those who know they are going to die. For that matter, we all know we are going to die; we just don't know exactly when. I think Jesus was offering a relief valve for such confusion stress when he said: "Do not let your hearts be troubled. Trust in God; trust also in me. In my Father's house are many rooms . . . . I am going there to prepare a place for you. . . . I will come back and take you to be with me" (John 14:1-3 NIV). On another occasion, Jesus offered a relief valve for everyday confusion stress: "[D]o not worry about your life, what you will eat or drink, or about your body, what you will wear. . . . Look at the birds of the air; they do not sow or reap or store away in barns, and yet your heavenly Father feeds them. Are you not much more valuable than they?" (Matthew 6:25-26).

A final relief valve for confusion stress pertaining to crises is another form of knowledge. It is good to know what to expect from yourself and from others as you and they undergo crises. According to scholars, there is a fairly predictable pattern that individuals experience during crises such as death and divorce. Kathleen Galvin and Bernard Brommel have taken what other scholars have observed and extracted four basic crisis stages that most people encountering crises go through.[60]

Stage one is the stage of "shock resulting in numbness or disbelief, denial." Technically speaking, this stage appears to be a natural relief valve. Just as your body goes into shock and numbness at the moment of a severe injury to provide you with at least temporary relief from traumatic corporal stress, so your mind provides a temporary relief valve from traumatic mental or confusion stress. Keep in mind, however, that this relief valve is only temporary.

Stage two is the "recoil stage resulting in anger, confusion, blaming, guilt, and bargaining." What is happening here is your stress balloon is exploding. The crisis has pumped entirely too much stress into your system. Your system has to release the stress suddenly. You explode in anger at God (blaming and bargaining), at the person who died (blaming), at the doctor(s) who could not save him/her (blaming), and even at yourself (blaming and guilt). If you face the crisis of divorce, you blame God, your minister or counselor for not saving the marriage, the partner who left, and yourself. If your parents were divorced, you blame them, God, any of your parents' new friends of the opposite sex. You need to realize, if you are going through a crisis that this stage will probably come. Try to anticipate your rage and direct it so that you do not lose friends in addition to those things you had previously lost.

Stage three is depression. Frankly, after your balloon explodes, you are in the state of no stress. Remember that this is not a good state. Human beings actually need some stress. It is called eustress. As mentioned earlier, no stress is usually the state of boredom. But, at stage three in going through a crisis, it is even worse. You are not just bored; you are totally depressed. You are apathetic. You may not even care whether you live or die. You may even believe that death may be preferable to this no stress state. You may attempt suicide. Understand what has happened to you. You have undergone a tidal wave of stress--the crisis. You survived initially through shock, but eventually you had to relieve your tremendous pressure and stress. You did that suddenly and dramatically in the recoil stage. But, since your balloon exploded (the stress was not just relieved slowly), you have zero stress left in you. You are depressed. You must begin to rebuild your stress levels to the eustress area. Having no stress is unacceptable.

---

[60] Sarah Trenholm and Arthur Jensen, *Interpersonal Communication* 2nd ed. (Belmont, CA: Wadsworth), 279.

The *fourth stage* is "reorganization resulting in acceptance and recovery." I find it interesting that crisis researchers use a form of the word "organization" to describe the final step toward recovery. As mentioned, organization is the key relief valve for confusion stress. Therefore, reorganization is the relief valve for the tremendous life-changing stress you have just undergone. You need to decide anew what your values and priorities are. You accept the fact that your life has changed and you begin to rebuild your organizational structure. Hopefully, you have not damaged friendships during the recoil stage. But, if you have, try to mend the relationships. Apologize. Yes, apologize. Explain to those you have offended that you were in the recoil stage of a crisis and that you did not handle it well. Rebuild relationships. Start back to work. Take on new challenges and pump new eustress into your balloon.

If you are aware of the four crisis stages, you should be able to weather even the most severe confusion stress. Remember, confusion stress results from being lost. If you know the roadmap for the territory you are entering, you may face many stresses, but at least you will know where you are at each stage.

Confusion stress has several relief valves. Knowledge is key. Know your own tendencies as a human when facing a crisis. Organize your life so you will know where you have put things. If something is confusing you, ask questions and keep asking until you know the answers. If you do not know the future but it looks bleak, be resourceful. Be willing to continually adjust your strategy with each new bit of knowledge that comes your way. Religious faith is a way of knowing the unknowable. Burke points out, for example, that Death itself is the Great Negative--it is not an "experience available to direct sensory knowledge."[61] Those who have religious faith can face much confusion calmly and relieve for themselves and others much confusion stress.

---

[61] Burke, *Language*, 455.

Chapter 8

Conscientious Stress

Ending the previous chapter by referring to religion makes a nice transition to this chapter. Conscience is thought to be a religious term. The apostle Paul uses the term in Romans 2:14-15 (NIV) to refer to those who were not familiar with the Hebrew religion, but who nevertheless might be successfully defended before his God: "Indeed, when Gentiles, who do not have the law, do by nature things required by the law, they are a law for themselves . . . since they show that the requirements of the law are written on their hearts, their consciences also bearing witness, and their thoughts now accusing, now defending them." Using Paul's definition, then, conscience implies following laws naturally, having the laws written on your heart.

While the term conscience is religious in origin, its application by Paul is without regard to religion. Paul is claiming that even non-religious individuals have consciences. Paul's non-religious contemporary, the Roman philosopher Seneca, made the observation that virtually all cultures dominated by the Romans agreed that humans should follow at least seven laws, which Seneca termed the seven "natural" laws.[62] The fact that Paul claims that non-religious people sometimes do "by nature" things that are in the "law," suggests that Paul may have accepted Seneca's observation and called this phenomenon "conscience."

Burke defines conscience as a "circl[ing] back upon the self" of those thou-shalt-nots you wish others to obey.[63] He asserts that the human penchant for morality (and the moral codes or laws that come with it) relates to the very nature of humans as symbol-users. Burke claims that humans are the symbol-using (symbol-making, symbol-misusing) animals. Implicit in Burke's claim is that no other animal communicates by the use of symbols. A symbol is an arbitrarily chosen sound, gesture, mark, light flash, electrical impulse, positive or negative charge, raised dot (as in Braille), and/or series and/or combination of the above to which humans have assigned meaning. The symbol represents or stands for something else. Symbol use only works if more than one person shares the meaning of a given symbol. For example, if I raise my right finger to my lips while making the sound "sh-h-h," most people know that I am communicating a request that they be very quiet. While other animals certainly communicate, they do not arbitrarily choose their own methods of communication. Bees appear to have an elaborate system of non-verbal communication signals, but this system is not arbitrarily chosen. So far as we can tell, bees from China can understand bees from South America. Bees are not free to invent their own systems of communication.

Humans, on the other hand, are free to invent new words, gestures, symbols for "Please be quiet." An elementary teacher may snap her fingers or raise two fingers or ring a bell to communicate the same message. This symbolic freedom (Burke understands it as free will) possessed by humans is the basis for Burke's observation that the human is *inventor of the negative (or moralized by the negative).*"[64] Humans are not only free to invent new symbols they can share; they are also free to invent laws and moral codes they can share. Their ability to invent these laws and codes is based upon their invention of what Burke calls the hortatory negative—the statement: Thou shalt not!

The Ten Commandments are largely comprised of hortatory negatives:

- Thou shalt not make any carved or sculpted idols.

---

[62] Seneca's laws may be phenomena that fit into a category of the unconscious described by Burke, *Language*, 72: "The universal incorporation of the past within the present."

[63] Burke, *Language*, 473.

[64] Burke, *Language*, 9-16.

- Thou shalt not use God's name in an empty manner.
- Thou shalt not kill.
- Thou shalt not commit adultery.
- Thou shalt not steal.
- Thou shalt not lie.
- Thou shalt not covet.

Furthermore, all human laws seem to be based upon hortatory negatives:

- Thou shalt not drive faster than 55 miles per hour on this highway.
- Thou shalt not discriminate on the basis of age, sex, religion, or national origin.
- Thou shalt not accept bribes in governmental positions.

Etiquette and manners are also based upon hortatory negatives:

- Don't chew with your mouth open.
- Don't belch in other people's presence.
- Don't wear a hat indoors if you are a man.
- Don't speak when someone else is speaking.

Clearly, all of these hortatory negatives imply freedom, free will, choice--the ability for humans to choose NOT to obey the specific hortatory negative. We only use a hortatory negative when it is clear that the one to whom the hortatory negative is addressed has a choice. If a mother tells her child not to play in the street, the child implicitly knows s/he has the power to disobey the mother's command. The collection or system of hortatory negatives that each individual human has constructed or received as valid is the basis of the individual's conscience. Whenever an individual is tempted to violate[65] (or actually violates) a hortatory negative s/he deems valid, that individual encounters conscientious stress. Generally, breaking a civil law (or being tempted to do so) produces more conscientious stress than does violating some point of etiquette (or being tempted to do so). As Burke understands the 18th century German philosopher Immanuel Kant, it is not enough to merely avoid violating thou-shalt-nots. There must be "consciousness of a *free* submission of the will to the [moral] law." Following moral law, according to Kant, involves the "thwarting of all our inclinations," including selfishness, and hence "produce[s] a feeling which may be called pain." Yet, it is also "something positive in itself." The "free are *free* to practice *self-*

---

[65] Burke, *Language*, 460, asserts: "[P]sychogenic illnesses would be infused with the 'Decalogical negative' insofar as they derive from repressions, inhibitions, self-punishments, and other judgments pronounced against the body by the mind it houses."

control." Kant sees that "'dignity' . . . arises from such a 'free submission of the will to the law' as amounts inevitably to 'constraint put upon the self.'"[66] I suggest that this "positive," which Kant calls "dignity," amounts to the feeling of relief one feels upon proving to oneself that one is free to practice self-control. Self-control is thus a relief valve for conscientious stress.

Violating a major religious commandment (or being tempted to do so) generally produces more conscientious stress for a religious believer than does violating a civic law or violating some point of etiquette. A young man and woman were members of a conservative church and accepted the hortatory negatives of that church including the following: Thou shalt not have an abortion. Although it had occurred nearly a decade earlier, the couple had produced a pregnancy outside of marriage that they had secretly aborted. Later, when they married, the woman became pregnant again, this time with twins. The babies were born but died soon after birth. The grief at the loss of their children was compounded by extreme conscientious stress. Each blamed himself/herself for the earlier violations of conscience. Unable to accept such guilt (conscientious stress), each blamed the other and exploded frequently toward each other with hate-filled speech. The marriage collapsed under the stress.

A retail salesman worked for a well-respected company. He believed his products were worth the price he asked customers to pay. He had no conscientious stress there. His company offered a satisfaction guarantee. Any customer who was dissatisfied with the performance of a product could demand either a replacement or a refund. He felt ethical about that--no stress. However, as customers began to invoke the quite liberal terms of the satisfaction guarantee, his superiors felt that the company could realize greater profits if they resisted customers' attempts to acquire replacements and refunds. The salesman was instructed by his superiors to argue vigorously with those customers who chose to invoke the guarantee. Generally, he was able to dissuade the elderly and the less aggressive from invoking the guarantee even though in his opinion they had legitimate claims. On the other hand, many whose claims he thought were frivolous were able to secure replacements and refunds by being very aggressive and obstinate. His superiors accepted this seemingly inequitable and unfair scenario so long as at least some of the claims were turned back. Nevertheless, the salesman encountered such conscientious stress from these procedures that he resigned his position with the company.

---

[66] Burke, *Language*, 441-442.

A businessman saw nothing wrong with the practice of pocketing cash from business transactions so that it would not be reported as income for tax purposes. He experienced no violation of conscience because this represented cash payments for services provided. There were no documents (checks, receipts, bills) that could be used by the IRS to verify that he had received any such payments. On the other hand, this same businessman considered cheating at golf (at times when no one else could see) to be the ultimate immoral act. Not everyone's conscience has a similar set of hortatory negatives.

On the morning of September 11, 2001, Americans awoke to a series of devastating events. Terrorists had hijacked four commercial airplanes. Two of the aircraft were flown loaded with passengers into the twin towers of the World Trade Center in New York. A third plane full of passengers was crashed into the Pentagon in Washington. The fourth dived to earth in Pennsylvania. This terrorist attack was the most devastating act of war on American soil since the Civil War. These terrorists apparently experienced no conscientious stress, however, from the prospect of murdering thousands of innocent civilians. Their consciences instructed them that suicide and murder in the proper cause were glorious, not reprehensible.

Timothy McVay was executed for his part in the Oklahoma City bombing. He refused to apologize even for killing so many children in the bombing. The children were just collateral damage, he claimed. The Nazi culture had no qualms about exterminating the Jewish people in World War II. High school students in the 1990s gleefully murdered their classmates at Columbine High School. In the 1960s, assassinations of prominent American figures were conducted without apparent conscientious stress. Conscientious stress is not synonymous with sin or crime. Many crimes (and sins) are committed without conscientious stress. When Disney's Jiminy Cricket instructs Pinocchio to "always let your conscience be your guide," he does not offer a panacea for society's ills. Conscience is a construct. Each individual's system of hortatory negatives is partially constructed out of personal experience and partially received from the various cultures with which that individual comes in contact. For the Branch Davidians to question David Koresh's teaching that they should take up arms against the Bureau of Alcohol, Tobacco, and Firearms (BATF) could have produced serious conscientious stress. Questioning Koresh was tantamount to atheism for Koresh's followers.[67] Much

---

[67] Stan A. Lindsay, "Waco and Andover: An Application of Kenneth Burke's Concept of Psychotic Entelechy" in *The Quarterly Journal of Speech* October 1999. See also Stan A. Lindsay, *Implicit Rhetoric: Kenneth Burke's Extension of Aristotle's Concept of Entelechy* (Lanham, MD: University Press of America, 1998) and Stan A.

more conscientious stress would have been produced by not firing at BATF agents than by firing.

Have No Moral Code

The purpose of this book, however, is not to instruct readers concerning a proper set of hortatory negatives. The objective is to help readers reduce conscientious stress. In line with that objective, it can be argued that one relief valve for conscientious stress is to have no moral code, no set of hortatory negatives. It is true that such a situation would produce no conscientious stress. However, such a situation is highly likely to produce high levels of community stress. Jesus summarized the numerous laws of the Bible in Matthew 7:12 (NIV): "So in everything, do to others what you would have them do to you, for this sums up the Law and the Prophets." Known as the Golden Rule, this ethical formula appears with slight alterations in virtually all cultures. If you treat others in ways you would not like to be treated, you will encounter community stress. Reciprocating, others will treat you in ways similar to your treatment of them. If you steal from them, they will steal from you. If you lie to them, they will lie to you. If you cheat on your spouse, your spouse will cheat on you or leave you. If you seek to kill others, others will seek to kill you. If you slander, you will be slandered. If you dishonor others, others will dishonor you. If you shout obscenities at others, they will be shouted back. Imagine the levels of community stress you will encounter if you have no moral code!

Businesses that use comparative advertising (advertising which compares one company's products with another company's products) might consider how much of the Golden Rule is operative in the following ten guidelines for comparative advertising suggested by the American Association of Advertising Agencies:

1. The intent and connotation of the ad should be to inform and never to discredit or unfairly attack competitors, competing products, or services.
2. When a competitive product is named, it should be one that exists in the marketplace as significant competition.
3. The competition should be fairly and properly identified but never in a manner or tone of voice that degrades the competitive product or service.
4. The advertising should compare related or similar properties or ingredients of the product, dimension to dimension, feature to feature.

Lindsay, *Psychotic Entelechy: The Dangers of Spiritual Gifts Theology* (Lanham, MD: University Press of America, 2005).

5. The identification should be for honest comparison purposes and not simply to upgrade by association.
6. If a competitive test is conducted, it should be done by an objective testing service.
7. In all cases the test should be supportive of all claims made in the advertising that are based on the test.
8. The advertising should never use partial results or stress insignificant differences to cause the consumer to draw an improper conclusion.
9. The property being compared should be significant in terms of value or usefulness of the product to the consumer.
10. Comparisons delivered through the use of testimonials should not imply that the testimonial is more than one individual's thought unless that individual represents a sample of the majority viewpoint.[68]

If you were the competitor whose product the advertised product was being compared with, wouldn't you like to be treated with the fairness these ten guidelines afford? Even though some community stress occurs in all competitive situations, that community stress is greatly reduced by adhering to a code of ethics based on the Golden Rule.

As witnessed by the corporal punishment doled out to Timothy McVey, Nazi war criminals, and others, it is also quite likely that you will encounter much more corporal stress if you choose to have no moral code. Corporal stresses as STDs result from cheating on your spouse. Corporal stresses from automobile accidents result from the road rage produced by nonverbal expressions of obscenity and disrespect. Even if diseases and counterattacks from those who have been offended do not produce corporal stress in someone who has no moral code, it is likely that the amoral person's own body will (unconsciously) translate the not-admitted conscientious stress into corporal stress. Burke writes: "[O]n many occasions . . . the violating of moralistic proscriptions may produce acute bodily distress and revulsion."[69]

Self-control

A second relief valve for conscientious stress is much more successful in reducing both conscientious and community stress. Diligently following the golden rule should make your life much less stressful. Of course, total perfection

---

[68] James B. Astrachan, "When to Name a Competitor," *Adweek* (May 23, 1988): 24
[69] Burke, *Language*, 423.

is fairly elusive. We all make mistakes, but the closer an individual approximates a perfect attempt to follow the golden rule, the less conscientious stress s/he experiences. There is very little guilt--the major conscientious stress. Others may not treat you the way you treat them, but at least you are not feeling guilty about the way you have treated them. Returning good deeds for evil ones and turning the other cheek when you are victimized may not rid you of corporal stress or community stress, but it will significantly reduce your conscientious stress. You will feel good that you have followed the dictates of your conscience even though others have not.

Seek Forgiveness

All humans violate their consciences at times. Feelings of guilt naturally result. What can be done to relieve the conscientious stress? It is called repentance. Take steps to correct the unethical behavior you have practiced. Apologize to any individuals who have been injured by your behavior. Consider ways you might make restitution. Have you cheated or stolen from someone? Find a way to pay them back--even if it takes a long time, even if you do it secretly. Your conscientious stress will be relieved. Have you slandered someone? Expend your efforts to correct the misinformation. Sing the praises of the person you have slandered to those to whom you have slandered that person. Burke equates cash stress and conscientious stress in this way: "Freedom from guilt, like freedom from debt or responsibility, can be felt as buoyancy." He distinguishes between "zero" and "minus" to explain: "Zero is the negating of any plus. . . . [M]inus is a super-negation . . . [that] in turn [needs] to be negated by a compensatory plus that brings us to zero."[70] Guilt (or conscientious stress) is like minus. There is the need for the guilty to provide a compensatory plus to bring himself/herself back to zero. Burke puts it succinctly: "'[P]ositive' *acts* can negate the guilt-negatives" so that there is "genius in penance, mortification, . . . ascetic regimens, [and] all rites."[71]

A thirty-year-old man faced a particularly difficult situation. He had been driving under the influence of alcohol and had caused an accident that resulted in the death of a young person. He repented. He stopped drinking and driving. He became active in church and sought forgiveness from God, but his conscience still bothered him every time the issue of drinking and driving came up. He did not

---

[70] Burke, *Language*, 435.
[71] Burke, *Language*, 459.

want to talk about the issue. How does one make restitution for having killed another human being? The answer may be the very thing he didn't want to do. Talk about it. If this man spent more of his time laying out before others the horrors he had experienced in being guilty for the death of another, some who heard him would have been persuaded not to drink and drive themselves. He will never be able to bring back to life the person he killed, but his personal story has the potential of saving many others. That might be a form of restitution.

Once you have apologized and asked for forgiveness and made restitution, you will find it easier to forgive yourself. When you are able to forgive yourself, your conscientious stress will be relieved.

In conclusion, although the relief valve of amorality (the option of having no moral code) was considered in this chapter, I do not recommend it. Much more community stress and corporal stress results from amorality. If the goal is to relieve overall stress, amorality is not pragmatic. Most religious and cultural codes are designed to actually assist the individual in relieving community and corporal stress. This goal is explicitly mentioned in the Torah (the first five books of the Bible) and in the New Testament. The fifth commandment of the Decalogue (Honor your parents) offers this compensation for following it: "that you may live long in the land the Lord your God is giving you" (Exodus 20:12 NIV). Deuteronomy 7:15 promises that following the Law will keep the nation free of many diseases. Mark 2:27 claims that the Law of Sabbath "was made for man, not man for the Sabbath" (NIV). These and other texts indicate that obeying the Judeo-Christian moral code will result in a lowering of Corporal and Community stress for the adherents.

Following a time-honored moral code to the best of your ability will probably result in lower conscientious stress. When you occasionally fail, pursuing forgiveness and repenting will lower conscientious stress providing a compensatory plus to bring yourself from a guilt-negative back to zero. As Burke puts it, "positive *acts* [of repentance] can negate the guilt-negatives" so that forgiveness can take place and conscientious stress can be lowered.

Chapter 9

The War on Terror: Every Stress

   After September 11, 2001, I began to write this final chapter on navigating the Seven Cs of Stress. On that date, America was faced with a multitude of stressors. The corporal stress endured by many (especially, citizens of New York and Washington, D.C.) has been vividly replayed many times in our minds and on our televisions. Terrorists, using small knives and cardboard cutters, slit the throats of flight attendants on commercial airlines as they lured pilot crews from their controls so the terrorists could overpower the pilots and take control of the planes. Upon taking control of the planes, the terrorists (some of whom had taken flight training in American schools) flew the planes loaded with fuel into the twin towers of the World Trade Center in New York City and the Pentagon in Washington. The impact instantly killed the terrorists, the passengers on board the planes, and those in the crash areas of the targeted buildings. Tremendous fires erupted, most notably in the twin towers. Many humans were either blown out of windows in the skyscraper towers due to the blast or jumped out of windows to avoid the intense heat of the fires. Others in the buildings fought through the smoke to stairwells, attempting to descend the many floors to the presumed safety of the street below or to ascend to the roof where they might be rescued by helicopters. As they descended the stairwells, those who had less corporal stress noticed others in wheelchairs who were, according to protocol, expected to wait in the stairwells until rescue workers could remove them. These surely felt not only more corporal stress but also more chrono stress as they awaited rescue.

We watched on our television screens as hundreds of rescue workers (police and firefighters) flooded toward the scene. These heroes rushed into an environment laden with corporal stress carrying their own share of chrono stress. How much time would they have to evacuate the survivors and extinguish the fires? Not nearly enough, as it turned out. Within minutes, the towers began to implode. One hundred stories per tower collapsed on one another. Dust, broken glass, stone, steel, debris, and smoke belched out of the towers and swept swiftly through the streets of New York. Those in the area (who were not even in the World Trade Center) felt the corporal stress of struggling for breath in the cloud of dust and suffered injuries as the debris pelted them. Many, including hundreds of rescue workers, were buried in the rubble. America faced a disastrous corporal stress. The only relief valve for this corporal stress for thousands was death itself. It came swiftly in most cases. At least, the corporal stress was not prolonged. For those who survived with injuries, our medical practitioners and facilities worked feverishly to alleviate stress. Many Americans also lined up at clinics all over the country to give blood.

Community stress rapidly ensued. Americans demanded to know who was responsible for these evil acts. Hatred swelled as the answer became apparent. Anti-American Arab Muslims had carried it out. Seeking immediate revenge, some Americans lashed out at any other Americans who resembled the perpetrators, either ethnically or religiously. An Egyptian Christian was murdered because he was an Arab. Moslem Americans who wore the traditional dress were targets of abuse. Even Americans of near eastern origin who were neither Arab nor Islamic were targets of hate crimes. It is true that when stress becomes intense distress, the balloon wants to burst, but there is no excuse for such indiscriminate and rash behavior. The evil of this behavior in attacking innocents is of the same variety as the evil of the terrorists. If someone feels such tremendous community stress that s/he must lash out physically at the enemy, an appropriate relief valve is enlistment in the armed services or obtaining a position in law enforcement. In such a context, it is much more likely that such a person's desire for physical revenge will be directed toward guilty rather than innocent parties.

While community stress increased in some relationships, it actually decreased dramatically in many others. For a period of time, many Americans stopped thinking of the Republican vs. the Democrat position on issues. There was just the American position. Latin Americans, African Americans, and Caucasian Americans closed ranks. Men and women united. Moslems, Jews, and Christians of all denominations joined in prayer. Even the majority of nations around the world that, at times, squabbled with the US expressed solidarity with America. The world in general was appalled by the evil acts of the terrorists. The world felt sympathy for America and vowed to help her fight terrorism. Many Islamic countries were among those that expressed solidarity, but not all. Some Islamic countries whose officials expressed solidarity witnessed some defiance and backlash from their own citizenry.

If we would relieve the community stress between Americans and some of the anti-Americans in the Moslem world, we should return to Chapter 3 and note that this is Cultural Stress not Psychological Stress. Furthermore, the specific Cultural Stress involved pertains to beliefs and values. Perhaps the biggest conflict in values American culture has with the perpetrators of this crime is in the area of religious pluralism. American culture was built on a foundation of religious toleration. In America, Moslems, Jews, Buddhists, and Christians of all varieties may practice their various religions without government interference. Even though many Christians believe Heaven is reserved for Christians, they do not typically engage in hate crimes against members of other religious parties (although, admittedly, it has happened from time to time). There are passages in the New Testament that have been used by some Christians as a basis for anti-Jewish attitudes and behaviors.[72] There are also passages in Jewish rabbinic literature that have been interpreted by Jews as anti-Christian. There is nothing in either Jewish or Christian authoritative scripture that is anti-Moslem because, of course, Mohammed was not born until long after the Jewish and Christian Bibles were codified. Yet (probably because the Koran was competing-for-acceptance with Judaism and Christianity), there are several anti-Jewish and anti-Christian comments in the Koran. By being ambiguous, the Koran makes the religious pluralism conflict even more difficult to resolve. While identifying Jews and Christians as "unbelievers" and "evildoers," the Koran recommends in some instances conducting holy war against unbelievers and evildoers.

---

[72] I have argued that the so-called anti-Jewish language of the New Testament was misunderstood and misused by anti-Semites. See Stan A. Lindsay, *Revelation: The Human Drama* (Bethlehem, PA: Lehigh University Press, 2001), 131-132.

Never once does the Koran explicitly command Muslims to fight and kill "Christians and Jews," but it is easy to see how some of these passages may be twisted into such a command. For a few years, I taught communication courses for Loyola University Chicago.   My classes contained many Muslims, Jews, Protestants, Buddhists, Hindus, and Roman Catholics.   No religious hatred or treachery was clearly visible from any of these groups.   Individuals with different religions can get along with each other.

America has defended persecuted Muslims in Kosovo and Bosnia. America has brokered peace agreements between Egypt and Israel. America defended Arab countries from their aggressive neighbor, Saddam Hussein. Although the U.S. is predominantly Christian in religion, America protects the rights of members of other religions to freely worship God in their own ways. America's ally, Israel, has shown good faith in its willingness to relinquish land to Egypt in exchange for secure peace agreements. Israel has shown it is willing to discuss similar matters with other Arab countries.

In the Koran, war is only permitted if the evildoers and unbelievers (not the believers) are the aggressors. Despite Saddam Hussein's protestations to the contrary, the United States was not the aggressor in the Gulf War. Hussein and his armies invaded Kuwait. America and several Islamic Arab countries, fighting together, repulsed the aggressor from Kuwait. The allies who removed Saddam from power in 2003, acted in response to Saddam's multiple violations of the agreement that had earlier left Saddam in power. America did not aggress against Bin Laden, either. He and his terrorists attacked the United States. Until the September 11, 2001 attack, the United States had only retaliated when attacked. The devastating terrorist attack on New York and Washington was totally unprovoked. Claiming the spirit of justice and self-defense advocated in the Koran, America had every right to defend itself from the aggressor. Most Moslem countries agreed somewhat, at least initially, and sided with the United States.

In addition to Corporal Stress and Community Stress, the attack on New York and Washington produced gigantic Cash Stress. The first week it was open after the terrorist attack, the New York Stock Exchange experienced its largest one-day and one-week drops in history. Of course, the cities of New York and Washington experienced cash stress due to the destruction of buildings and the clean-up expense. The airlines industry, which was already experiencing considerable cash stress prior to the attack, was financially devastated by the resulting reduction in ridership.   Many companies laid off workers.   Talk of a coming recession reverberated in financial circles.

In terms of relief valves, the companies that laid off workers were attempting to use the "decrease expenses" relief valve. However, on a national scale, the best relief valve appeared to be to "increase income." President Bush and the Congress promptly passed a forty billion dollar aid package for New York and Washington. They also put together a bailout plan for the airline industry. Prior to the attack, they had passed a tax relief package, which funneled surplus tax dollars back into the economy. Throughout America, collections were taken in churches, synagogues, and mosques, on street corners, online, and by phone. Americans pumped millions of dollars into rescue and aid packages for those who were hardest hit. America served as a real-life version of the Christmas movie classic, "It's a Wonderful Life."

Chrono stress had affected rescue workers before the towers collapsed in New York; there was not enough time to complete their tasks. Many in the World Trade Center and many of their would-be saviors perished when the towers collapsed. Others, however, did escape. The evacuation of the towers was swift and orderly. Otherwise, the death toll could easily have doubled or tripled. Schools in the area were evacuated in a rapid, orderly fashion because those in charge had specific evacuation plans. Tons of debris were rapidly and carefully removed over the next week as rescue workers came in shifts trying to locate possible survivors under the rubble. Only a handful of survivors were found in the rubble, but every life was valuable as the rescuers fought against the clock.

Competence stress may have been present in the minds of the leaders of our country, but you could not tell it. Our nation's security plans were rapidly employed. The White House and the Capitol building were evacuated at various times. The President was moved from one place to another to prevent a possible assassination. Even as the President and (third-in-line-to-the-Presidency) Speaker of the House were present as the President addressed Congress, the Vice-President and fourth-in-line-to-the-Presidency were kept secure at another location in case the unthinkable occurred. The Secretary of State and other members of the President's cabinet made contacts with the heads of other nations to coordinate America's response. The intelligence services went immediately to work sifting through the evidence. America was prepared.

America faced a great deal of Confusion Stress. The FAA did not know if it was safe to fly, so the agency shut down all air travel except that of the military. When the agency reopened the airways to traffic, new and improved security rules were put in place. Even so, rumors of potential new terrorist attacks caused many travel plans to be postponed. Walt Disney closed its theme parks. Major League

Baseball games, NFL and NCAA football games, and many other high-profile, highly attended events were cancelled until confidence that they would be safe could be regained. Americans were glued to their TV sets doing precisely what should be done when people are confused--getting questions answered. The government and the media were very forth coming with the information America needed.

Except for the occasional bigot who wrongly attacked innocent people of near eastern descent and the occasional scam artist seeking contributions for bogus charities, there was little conscientious stress in America in the immediate aftermath of the attack. America stood tall. In a spirit of national brotherhood, Americans in unison treated their fellow citizens in the way they would want to be treated. In San Francisco, a few years earlier, the phenomenon had been termed "earthquake love." Houses of Worship were filled. Blood banks had to turn donors away. Millions of dollars were donated to charity.

Three years later, as I complete the editing of this book, America has still not been hit again by a major terrorist attack. America's response to 911 has been effective, so far. Despite a very intense political campaign, both candidates for president agree with the American citizens that the war on terror is the number one priority. I suppose there will be those who read this book who will think that the stress they personally face is too great to navigate. I hope such readers find in the example of America's response to the extreme stress of a terrorist attack proof that they can personally navigate any stresses known to humanity.

The method I propose is very simple:

1. Diagnose the stressor (determine which of the seven Cs is responsible for your stress). The Seven Cs are corporal stress, community stress, cash stress, chrono stress, competence stress, confusion stress, and conscientious stress. A mnemonic device for remembering the seven is the following: Two Cs begin with "com" (community and competence). Two Cs begin with "con" (confusion and conscientious). One C begins with "cor" (corporal). The other two are cash and chrono, which are easily remembered by the proverb, "Time is money" (or, put in terms of Cs, "Chrono is cash). So "com, con, cor; time is money" should help you remember the seven Cs.

2. Seek out and apply the appropriate relief valves:
   - Under corporal stress (any stress that affects the body), there are two types: translated corporal stress and originating corporal stress. Translated corporal stress may have its origin in any of the other 6 Cs. Virtually all stress (regardless of which of the stressors is responsible

for it) eventually becomes translated into corporal stress. The appropriate relief valve for translated corporal stress is to determine which of the other stressors caused it and to employ the relief valves of the originating stressor. Originating corporal stress is the stress that actually begins somewhere in the body. All bodily systems should be considered—cardiovascular, respiratory, nervous, reproductive, musculoskeletal, integumentary, endocrine, etc. Each bodily system has its own relief valves. Additionally, taking protective measures and employing the universal relief valves of laughter (comedy) and tears (tragedy) will help to relieve the stress.

- Community stress is the stress you feel when you must get along with other human beings. It can be divided into psychological stress and cultural stress. Psychology refers to the behaviors, values, and characteristics of an individual. Culture refers to the behaviors, values, and characteristics of a distinct group of people. Relief valves for psychological community stress are approaches such as democracy, tyranny, and anarchy. Relief valves for cultural stress are as follows: Study the cultures of those with whom you come in contact. Decide which values you hold that require cross-cultural unity and debate them. Seek out other cultures that might agree with you on those essential values and build relationships/coalitions. Identify those ways of doing things and ways of behaving that are purely arbitrary, sheer matters of opinion and allow for plurality and anarchy in these areas. Rely frequently on your own culture to provide an environment where the community stress is lessened. Consider those areas in which your culture holds values that other cultures do not hold. Will society collapse if only your culture holds these values? If not, encourage others in your culture to keep them; do not demand that all other cultures change.

- Cash stress is either the stress you feel when you do not have enough cash to meet your wants or the stress you feel when you have plenty of cash and are not sure how you should invest it. If you have the first type, it is the fault of someone else, the fault of no one, or your fault. If someone else is at fault, you may want to avoid such situations. If you are at fault because the stress was preventable, you might consider buying insurance to guard against the cash stress possibility. If no one is at fault, but the cash stress affects large

- numbers of individuals, you may find relief in what Burke terms the socialization of losses. The two primary relief valves for cash stress are increasing your income and decreasing your expenses. Financial planning is the process used to accomplish all of these forms of relief.

- Chrono stress is the stress you feel when you have too much to do and too little time in which to do it. The relief valve, time management, includes four steps: List tasks to be performed. Prioritize these tasks. Allocate time for performing each task. Schedule the tasks, beginning with the highest priorities.

- Competence stress is the stress you feel when you consider yourself incompetent to perform a task you must do. The primary relief valves are practice, preparation, and receiving signs of approval. Additionally, it is wise to choose your own hierarchies.

- Confusion stress is the stress you feel when you are lost. There are several relief valves. Knowledge is key. Know your own tendencies. Organize your life. Ask questions. Have a religious faith.

- Conscientious stress is the stress you feel when you are tempted to violate one of your "thou shalt nots." Following a time-honored moral code to the best of your ability will probably result in lower conscientious stress. When you occasionally fail, pursuing forgiveness and repenting will lower conscientious stress.

3. Remember that your goal is to control your level of stress, not eliminate it completely.

This book is filled with anecdotes, scenarios, and case studies. These are stories designed to help you understand how stress operates, but they can serve another purpose. Just as tragedy helps purge fear and pride from an audience through tears shed in pity for some tragic victim, so these stories can purge stress from readers who identify with the situations described in the stories. Burke sees relationships between "nation, class, family, [and] individual."[73] Though a given story may demonstrate a stress between nations, readers may be able to translate the story easily into a stressful situation involving labor and management or a stressful situation within their own families or between families or individuals.

One reason these stories may be useful in reducing community stress is that the story provides a certain "distance." Burke writes: "[A] certain 'distance' could be got in Shakespeare's day by treating the [contemporary social] problem in terms

---

[73] Burke, *Language*, 90.

not of contemporary London but of ancient Rome."[74] Similarly, by relating the stories in this book to others, you may be able to provide enough distance that the community stresses may be solved by encouraging others to look into the mirror of similar stories/situations.

It even helps that some of the stories add stress to the situation. Burke suggests that instead of "toning down such situations, the dramatist must work his cures by . . . find[ing] ways to *play them up*."[75] If there is more community stress in one of the stories offered here (or elsewhere) than you find in your own community stress situation, the successful resolution of the story you offer may serve to minimize or relieve the stress of your actual situation. Now, I wish you Bon Voyage as you successfully navigate your own versions of the Seven Cs of Stress!

---

[74] Burke, *Language*, 82.
[75] Burke, *Language*, 82.

Bibliography

Astrachan, James B. "When to Name a Competitor" *Adweek* (May 23, 1988): 24

Berko, Roy M., Rosenfeld, Lawrence B., and Samovar, Larry A. *Connecting: A Culture-Sensitive Approach to Interpersonal Communication Competency,* 2nd ed. Ft. Worth, TX: Harcourt Brace College Publishers, 1997.

Wayne C. Booth, *Modern Dogma and the Rhetoric of Assent* Chicago: Univ. of Chicago Press, 1974.

Burke, Kenneth. *Attitudes Toward History*. 3rd ed. Berkeley: Univ. of California Press, 1984.

---. *Attitudes Toward History*. 2 vols. New York: New Republic, 1937.

---. "Bodies That Learn Language." Lecture at University of California, San Diego, 1977.

---. *The Complete White Oxen*. Berkeley: Univ. of California Press, 1968.

---. "Counter-Gridlock: An Interview with Kenneth Burke*" All Area*, vol 2 (Spring, 1983): 4-31.

---. *Counter-Statement*. Berkeley: Univ. of California Press, 1968.

---. "Dramatism." In *Communication: Concepts and Perspectives*. Edited by Lee Thayer. 327-360. Washington, DC: Spartan Books, 1967.

---. *Dramatism and Development*. Barre, MA: Clark Univ. Press with Barre Publishers, 1972.

---. "The Five Master Terms: Their Place in a 'Dramatistic' Grammar of Motives." *View* 3, no. 2 (1943): 50-52.

---. "Freedom and Authority in the Realm of the Poetic Imagination." In *Freedom and Authority in Our Time*. Edited by Lyman Bryson, Louis Finkelstein, R. M. MacIver, and Richard McKeon. 365-375. New York and London: Harper & Brothers, 1953.

---. *A Grammar of Motives*. Berkeley: Univ. of California Press, 1969.

---. *Language as Symbolic Action: Essays on Life, Literature, and Method*. Berkeley: Univ. of California Press, 1966.

---. "On Catharsis or Resolution, with a Postscript." *Kenyon Review* 21 (1959): 337-375.

---. "On Human Behavior Considered 'Dramatistically.'" In *Permanence and Change: An Anatomy of Purpose*. 2nd ed. 274-294. Indianapolis: The Bobbs-Merrill Company, Inc., 1975.

---. "On Stress, Its Seeking" in *Why Man Takes Chances: Studies in Stress Seeking*. Edited by Samuel Z. Klausner. 75-103. Garden City, NY: Doubleday, 1968.

---. *On Symbols and Society*. Edited by Joseph R. Gusfield. Chicago and London: Univ. of Chicago Press, 1989.

---. "Othello—An Essay to Illustrate a Method." In *Perspectives by Incongruity*. Edited by Stanley Edgar Hyman. 152-195. Bloomington: Indiana Univ. Press, 1964.

---. *Permanence and Change: An Anatomy of Purpose*. 2nd ed. Indianapolis: The Bobbs-Merrill Company, Inc., 1975.

---. *The Philosophy of Literary Form: Studies in Symbolic Action*. 3rd ed. Berkeley: Univ. of California Press, 1973.

---. "Poetics and Communication." In *Perspectives in Education, Religion, and the Arts*. Edited by Howard E Kiefer and Milton K. Munitz. 401-418. Albany: State Univ. of New York Press, 1970.

---. "Questions and Answers about the Pentad." *College Composition and Communication* 29 (1978): 330-335.

---. *A Rhetoric of Motives*. Berkeley: Univ. of California Press, 1969.

---. *The Rhetoric of Religion*. Boston: Beacon Press, 1961.

---. "Rhetoric—Old and New." In *New Rhetorics*. Edited by Martin Steinmann, Jr. 59-76. New York: Scribner's Sons, 1967.

---. "Rhetoric, Poetics, and Philosophy." In *Rhetoric, Philosophy, and Literature*. Edited by Don M. Burks. 15-33. West Lafayette, IN: Purdue Univ. Press, 1978.

---. "The Rhetorical Situation." In *Communication: Ethical and Moral Issues*. Edited by Lee Thayer. 263-275. London, New York, Paris: Gordon and Breach Science Publishers, 1973.

---. "Tactics of Motivation*" Chimera*, vol 1 (1943): 27-44.

---. "Theology & Logology*" Kenyon Review*, New Series, vol I # 1, Winter 1979.

---. *Towards a Better Life*. Berkeley: Univ. of California Press, 1982.

DeVito, Joseph. *The Elements of Public Speaking* 6th ed. Longman Publishing Co.

Greenberg, Susan H. and Springen, Karen. "The Baby Blues and Beyond" *Newsweek* (July 2, 2001): 26.

Hager, W. David and Hager, Linda Carruth. *Stress and the Woman's Body*. Grand Rapids, MI: 1995.

"Infanticide: A History" *Newsweek* (July 2, 2001): 22-23.

Jay, Paul, ed. The Selected Correspondence of Kenneth Burke and Malcolm Cowley 1915-1981. Berkeley and Los Angeles: Univ. of California Press, 1990.

Jennermann, Donald L. "Kenneth Burke's Poetics of Catharsis." In *Representing Kenneth Burke*. Edited by Hayden White and Margaret Brose. 31-51. Baltimore and London: John Hopkins Univ. Press, 1982.

Klausner, Samuel Z., ed. *Why Man Takes Chances: Studies in Stress-Seeking* Garden City, NY: Doubleday, 1968.

Lindsay, Stan A. *Implicit rhetoric: Kenneth Burke's Extension of Aristotle's Concept of Entelechy* Lanham, MD: Univ. Press of America, 1998.

---. *Persuasion, Proposals, and Public Speaking*. Orlando: Say Press, 2004.

---. *Psychotic Entelechy: The Dangers of "Spiritual Gift" Theology*. Lanham, MD: Univ. Press of America, 2005.

---. *Revelation: The Human Drama*. Bethlehem, PA: Lehigh Univ. Press, 2001.

---. *The Twenty-One Sales in a Sale*. Grants Pass, OR: Oasis Books/PSI Research, 1998.

---. "Waco and Andover: An Application of Kenneth Burke's Concept of Psychotic Entelechy" *Quarterly Journal of Speech*, vol 85 (1999): 268-284.

McKeon, Richard. *Introduction to Aristotle*. 2nd ed. Chicago and London: Univ. of Chicago Press, 1973.

Norheim, Karen. "A Heart for God" *The Lookout* (January 29, 1989): 8.

O'Hair, Dan, Friedrich, Gustav W., Wiemann, John M., and Wiemann, Mary O. *Competent Communication*, 2nd ed. New York: St. Martin's Press, 1997.

Pomfret, John. "In China's Countryside, 'It's a Boy!' Too Often" *The Washington Post* (May 29, 2001): A01.

Samovar, Larry A. and Porter, Richard. *Communication between Cultures* Belmont, CA: Wadsworth, 1991.

Smith, Robert E. *Principles of Human Communication* 4th ed. Dubuque, IA: Kendall/Hunt, 1995.

Sprague, Jo and Stuart, Douglas. *The Speaker's Handbook* 5th ed. Fort Worth: Harcourt College Publishers, 2000.

Trenholm, Sarah and Jensen, Arthur. *Interpersonal Communication* 2nd ed. Belmont, CA: Wadsworth.

Index